# Rethinking IT in Construction and Engineering

How can the potential of IT be realised to improve business performance in architecture, construction and engineering organisations?

How can organisations unleash the potential of IT to achieve a sustainable competitive advantage?

How to migrate from technology to IT-enabled business thinking?

Based on the author's 20 years research experience in this field, this book provides a holistic picture of the factors that enable architecture, construction and engineering organisations to explore the potential of IT to improve their businesses and achieve a sustainable competitive advantage. It raises awareness of the importance of the organisational "soft issues" and the role they play in influencing the outcome of IT investments. The book also addresses other complementary enablers such as knowledge management, learning organisations, maturity models and e-readiness measurements.

Real case studies are used throughout the book to illustrate various concepts and to provide the reader with a realistic and practical picture. *Rethinking IT in Construction and Engineering: Organisational Readiness* is aimed at lecturers and researchers in architecture, construction and engineering. It will also be relevant to professionals at managerial level in industry.

**Mustafa Alshawi** is the Director of the Research Institute for the Built and Human Environment, University of Salford, UK. He is the Editor-in-Chief of the international journal *Construction Innovation*.

Also available from Taylor & Francis

# Rethinking IT in Construction and Engineering

## Organisational readiness

Mustafa Alshawi

Routledge
Taylor & Francis Group

LONDON AND NEW YORK

First published 2007
by Taylor & Francis

2 Park Square, Milton Park, Abingdon, Oxfordshire OX14 4RN
52 Vanderbilt Avenue, New York, NY 10017

First issued in paperback 2020
First issued in hardback 2019

*Routledge is an imprint of the Taylor & Francis Group,*
*an Informa business*

Copyright © 2007 Mustafa Alshawi

Typeset in Sabon by
RefineCatch Limited, Bungay, Suffolk

*British Library Cataloguing in Publication Data*
A catalogue record for this book is available from the British Library

*Library of Congress Cataloging-in-Publication Data*
Alshawi, Mustafa.
    Rethinking IT in construction and engineering : organisational
    readiness / Mustafa Alshawi.
       p. cm.
    Includes bibliographical references and index.
    ISBN 978-0-415-43053-1 (hardback : alk. paper) 1. Building–
Data processing. 2. Construction industry–Data processing.
3. Communication in the building trades. 4. Preparedness.
5. Organizational change. I. Title.
    TH437.A465 2007
    624.0285–dc22                                    2007000830

ISBN 13: 978-0-367-57765-0 (pbk)
ISBN 13: 978-0-415-43053-1 (hbk)

# Contents

# Figures

# Tables

# Preface

There have been significant advances in the development and application of information technology (IT), both in hardware and software. Such advances have clearly influenced the way in which organisations have integrated IT into their business environment. It is widely accepted that IT is becoming a key element of any organisational infrastructure. Indeed, many like to think that the level of an organisation's reliance on IT in the twenty-first century is similar to the reliance on electricity in the previous century where it was not expected for an organisation to function without electricity. For example, networks, Internet, e-mail and office automation are seen as standard applications for "reasonable" size organisations. For small businesses, stand-alone applications such as e-mail, presentations and report writing are seen to be essential components for running any business. However, for larger organisations, the picture is more complex where IT infrastructure plays a key role in supporting core business functions. In this context, IT is being increasingly used to support business strategies as an enabler to leverage its potential to gain a competitive advantage and therefore new markets/clients.

However, there is ample evidence that information systems (IS)/IT have failed to bring about a competitive advantage to organisations in spite of the large investments over the past decade. A large percentage of IS/IT systems have failed to achieve their intended business objectives. Previous studies in the area of "IS/IT failure" have shown that 80 to 90 per cent of IT investments did not meet their performance objectives. Such projects were abandoned, significantly redirected, or even worse, they were "kept alive" in spite of their failure. The cost of funding such projects and the missed opportunities of not benefiting from their intended capabilities constituted a tremendous loss for organisations. This dissolution in the strategic benefits of IS/IT is currently forcing many organisations not to invest in IT for any competitive advantage but for the reasons of bringing efficiency and effectiveness to business processes.

The main attributes of the high percentage of systems' failure are rarely purely technical in origin. They are more related to the organisational "soft issues" which underpin the capability of the organisation to successfully absorb IS/IT into its work practices. IT is still, in many cases, being considered by the management of organisations as a cost-cutting tool (owned and managed by their IT departments). This "technology push" alone – even though to some extent it is still dominating in many industries like construction and engineering – will not harness the full business potential of IS/IT and thus is unable to lead to competitive advantage. Although the implementation of a few advanced IT applications might bring about "first comer" advantage to an organisation, this will not last long as it can be easily copied by competitors. It is innovation in process improvement and management, along with IT as an enabler, that is the only mechanism to ensure sustainable competitive advantage. This requires an organisation to be in a state of readiness, which will give it the capability to positively absorb IS/IT enabled innovation and business improvement into its work practices.

The competencies that an organisation needs to develop in order to acquire the capability to strategically benefit from IS/IT, prior to IS/IT investment, involve four main elements: people, process, work environment and IT infrastructure. The first two elements are the key to change and improvement while the other two elements are enablers without which the first two elements cannot be sustained. The "acceptable" level of IS/IT, that can be successfully utilised in an organisation – i.e. ensuring its business benefits are realised – therefore depends on assessing a range of critical issues needed to ensure a balance between the organisation's readiness (mainly factors required to adapt to the proposed change) against the level and complexity of the proposed IT. This balance often includes many issues such as capital expenditure, resource availability, an organisation's maturity and readiness, culture and vision, and available IS/IT skills.

In order for an organisation to achieve the required level of capability to address IS/IT based innovation and continuous improvement, the following will be necessary.

(a) *Create an innovative work environment.* This should focus on developing and sustaining a highly skilled and flexible workforce which will have the skills and the competencies to continuously introduce improvement through better and more streamlined business processes enabled by advanced IT. In this context, organisational learning and knowledge management become a necessity for organisations to sustain business improvements and competitive advantage out of their IS/IT investments.

(b) *Achieve effective alignment between business strategies and IS/IT strategies.* The focus should be on improving the organisation's efficiency by directly integrating IS/IT with the corporate, strategic and operational needs. This ensures IS/IT resources are "in line" with business imperatives.

To develop an IS/IT capability, an organisation needs to rethink its processes, structure and work environment. This necessitates the development of a "forward looking" management tool which will enable managers to:

1    measure their current capabilities in the relevant areas, i.e. those that affect the development of the required IS/IT capabilities
2    predict the required level of change and resources to develop the target capabilities, i.e. identifying the organisation's "readiness gap" for developing and adopting specific IS/IT capabilities.

This book focuses on the organisational soft issues that hinder the successful implementation of IS/IT in an attempt to raise the level of understanding of these issues and to highlight how best to deal with such issues through IS/IT readiness measurement tools. Real case studies are used throughout the book to illustrate various concepts and provide readers with a realistic and practical picture.

The content of the book is structured in three main parts.

## Part 1 Elements of IS/IT success and measurement in organisations

This part (Chapters 1–4) of the book sets the scene by clearly explaining the trends in the development, management and success measurement of IS/IT projects along with their uptake and impact on business. It then examines the technology impact on the construction and engineering industry. Chapter 1 addresses the pace of change in hardware and software along with the implementation patterns and problems that this industry has gone through over the past decade in order to improve its performance and efficiency. It also discusses the evolving role of IS/IT management in organisations and relates that to the success of IS/IT. It also covers examples of successful applications of IS/IT in the Construction and Engineering industry. Chapter 2 focuses on the relationship between the implementation of IS/IT and the dynamics of businesses within organisations. It clearly explains the relationship between the success of IS/IT projects and the capability of organisations to successfully absorb new systems within their work practices. The focus of the discussion is on the four organisational elements of success: people, process, work environment and IT infrastructure. Chapter 3 is mainly concerned with IS success measurement models. It focuses on product-based measurement approaches and classifies them into two categories, hard and soft. Chapter 4 illustrates the context of this book in two detailed case studies. The chapter emphasises the role of organisational soft issues and their impact on the success or failure of IS/IT projects.

## Part 2 Enablers: learning organisations and strategies alignment

This part of the book (Chapters 5–7) addresses the organisational enablers which are not only necessary to foster IS/IT capabilities in organisations but also ensure sustainable IS/IT competitive advantage. The focus of this part is on two issues: a) organisational learning/knowledge management as being the catalyst for knowledge sharing and capacity building and b) integrating business and IS/IT strategies. Chapter 5 explains and discusses the concept of learning organisation and the role of knowledge management as a tool to achieve continuous improvement. The concept of learning is linked to the resource-based models in an attempt to explain how such an environment can be achieved. Chapter 6 addresses the importance of developing holistic implementation strategies for developing learning environments and uses knowledge management as a case. A three layer holistic implementation model is presented which covers the entire knowledge management cycle. Chapter 7 is concerned with the alignment of IS/IT strategies with business strategy. It provides backgrounds to the IS, IT and business strategies followed by a high level implementation framework. The chapter ends with a focus on IT training, as a key concept to building capability.

## Part 3  Improvement gaps and tools: maturity and readiness of organisations

This part (Chapters 8–10) focuses on building IS/IT capabilities of organisations and measuring their readiness gap prior to IS/IT investments. Maturity concepts and models are used to categorise the organisational elements of success (process, people, work environment, IT infrastructure) in several levels of maturity which are then used to identify the current and required level of organisational readiness in the four elements of success. An IS/IT readiness model is proposed to address this. This concept is then discussed within the IS life cycle where a new stage "Readiness Gap" is proposed to the IS life cycle before the development and implementation phase starts. Chapter 8 is concerned with maturity and maturity models. It uses a number of application examples to illustrate the concept. In addition, this chapter briefly covers the various process improvement models in the software industry where concepts are adopted in the proposed readiness models. Chapter 9 addresses the definition of opportunity and readiness gap and attempts to modify the IS life cycle by adding a new stage to measure the IS/IT capability of organisations prior to the commencement of the development and implementation phase. The chapter ends with a detailed IT training strategy as an example of the preparation of IS/IT skills in organisations. Chapter 10 presents the proposed IS/IT readiness model and utilises the full case studies, explained in Chapter 4, to illustrate the implementation of the model.

# Acknowledgements

The content of this book is based on the author's many years of extensive research experience in this field. His work at the University of Salford (UK) has focused on organisational soft issues within the context of the construction and engineering industry. Parts of the book are based on the work of four Ph.D. graduates who completed their dissertations under the supervision of the author. These colleagues and friends, to whom I am grateful for their achievements and contributions, are Dr Yasser Salah (Ph.D. title "IS/IT Success and Evaluation: A General Practitioner Model"), Dr Jack Goulding (Ph.D. title "IT Training Strategies for the UK Construction Industry"), Dr Ahmed Obaid (Ph.D. title "Model for a Successful Implementation of Knowledge Management in Engineering Organisations"), and Mr Hafez Salleh (Ph.D. title "Measuring the IS/IT Readiness of Organisation: An IS/IT Maturity Model").

Figure 5.4 from "Organisational Learning and Core Capabilities Development: The Role of IT", by C. Ciborra and R. Andreu in R. Galliers and W. Baets (eds) *Information Technology and Organisational Transformation: Innovation for the 21st Century Organisation*, 1998. Copyright John Wiley & Sons Limited. Reproduced with permission.

Figure 5.5 reprinted from J. Peppard and J. Ward, "Beyond Strategic Information Systems: Towards an IS Capability" *The Journal of Strategic Information Systems*, 2004, 13(2): pp. 167–194. Copyright (2004), with permission from Elsevier.

Figures 7.3, 7.5, 7.11 from J. Ward and P. Griffiths, *Strategic Planning for Information Systems*, 1997. Copyright John Wiley & Sons Limited. Reproduced with permission.

Figures 7.6 and 7.7 from M. Alshawi and G. Aouad "A Framework for Integrating Business and Information Technology in Construction", *Civil Engineering and Environmental Systems*, 1995, 12: pp. 249–261, http://www.informaworld.com.

Figure 8.1 from Construct IT 2000.

Figure 8.2 reprinted from K. Layne and J. Lee, "Developing Fully Functional E-government: A Four Stage Model", *Government Information Quarterly*, 18: pp. 122–136. Copyright (2001), with permission from Elsevier.

Figure 8.3 from M. Taylor, *E-BC Strategic Plan, Performance Measures*, version 3, 2003, State of British Columbia, http://www.cio.gov.bc.ca/ebc/discussion/Performance_Measures_Report_Final.pdf.

Special permission to use Figure 8.4, "The Key Process Areas by Maturity Level" from "Capability Maturity Model for Software, Version 1.1.", Technical Report CMU/SEI–93-TR_024, Copyright 1993 Carnegie Mellon University, and Figure 8.5, "Process Areas of the People CMM®" from "People Capability Maturity Model® (P-CMM) Version 2.0", Maturity Model CMU/SEI–2001-MM–001, Copyright 2001 Carnegie Mellon University is granted by the Software Engineering Institute.

# Part I

# Elements of IS/IT success and measurement in organisations

Chapter 1

# Current status of IS/IT management and applications

## 1.1 Advances in hardware and software

There have been significant advances in the development and application of information technology (IT) over the past decade. Computers have moved from mainframes in the 1960s and 1970s to personal computers in the 1980s and 1990s. This was accompanied by a rapid development of networks (local area networks and wide area networks) which facilitated the transfer and exchange of information between personal computers within an organisation and among organisations. The emergence of the Internet in the 1990s added another significant dimension to the IT industry. This global network is facilitated by significant advances in communication protocols and standards along with the establishment of enormous optical cable networks. This started with the telephone as a medium for transferring and exchanging information between computers to broadband networks using optical fibres. Such a development, in the hardware and networks, had a great influence on the speed of processing information not only within a single computer but also on the speed of transferring information from one computer to another through optical and wireless networks.

Advances in hardware and networks were accompanied by a significant development in the software industry. Moving from machine coding and Cobol as a programming language to four generation languages, artificial intelligence, CASE tools and Java. Better and more powerful computer applications were developed covering a wide range of business and professional applications. This development has clearly influenced the way in which organisations have adopted IT into their business environment. In the 1970s and the 1980s, where the availability of commercialised application packages was limited, organisations developed and maintained their own specific business software applications. The complexity and high cost of developing bespoke systems led to the creation of a generation of business-oriented vendors in the 1990s. Powerful, reliable and cost-effective commercial professional and business application packages started to appear in the market. Hence, the adoption of IT applications started to shift from organisations developing their own application packages to subscribing to application service providers (ASPs) or buying a licence for a commercially available business package from a third party. More recently, organisations have started to change their strategies in favour of out-sourcing part, if not all, of their IT applications (non-core business functions) to a third party with the aim of achieving efficiency and cost effectiveness.

## 1.2 Patterns of IT focus in construction and engineering

The type and nature of IT investments and applications in the construction and engineering industry, over the past 10 to 15 years, have taken several shapes and followed many themes which can be categorised into four phases. Each phase has specific features which generated a "cloud" of IT interests among managers in both the IT sector and industry (see Figure 1.1). These four phases are IT infrastructure, process-focus, supply chain-focus and Internet-focus.

### Phase I: IT infrastructure

In the 1990s, organisations worldwide invested heavily in IT with the aim of improving business performance, gaining efficiency and a competitive advantage. Investment in IS/IT in 1996 was $11 trillion of which $250 billion was invested in the USA and $35 billion in the UK (Willcocks and Lester 1996). The majority of this investment went on setting up the IT infrastructure in organisations including hardware, networks and business applications. Although the amount of computing power per worker grew in organisations, researchers found that the productivity in those organisations did not reflect it. Also there was no clear evidence that such investments led to reaping the benefits of bottom line organisational performance improvement (Brynjolfsson and Hitt 1998; Brown 1987; Roach 1987). A worldwide survey of senior executives, carried out by the London Business School in 1999, showed that 80 per cent of chief executives were disappointed by the contribution of IT investments to business perform-ance. However the same survey indicated that more than 40 per cent of chief executives believed that the Internet will transform their business and that more than 50 per cent predicted that within three years the Internet will have a major impact on their businesses (*Economist* 1999).

### Phase II: process focus

The large investments in the 1990s were mainly propagated by advances in IT which has led to technology focused solutions to business problems. It became clear that by applying IT to existing work practices, IT only managed to speed up the time for information to travel between process bottlenecks, without doing anything to mitigate the effect of the bottleneck in the first place. As a result, organisations started in the mid 1990s to lean towards improving

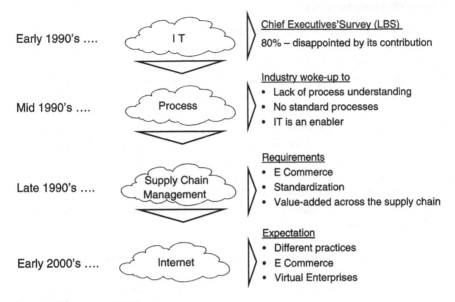

*Figure 1.1* Patterns of IT focus.

business processes as the main driver to improve business performance. By doing so, construction and engineering industry started to recognise the need for process understanding and process standardisation (see Case study 2.1), and above all that IT is only an enabling tool to the implementation of better and more effective business processes (see Case study 2.2).

### Phase III: supply chain focus

In the late 1990s, there was a shift from organisation-centred improvement, i.e. utilising IT as a tool to improve business performance, to better managing the supply chain process with the aim of not only adding value to "own" business but also to adding value to client organisations. This shift was also driven by pressure for improvements which started to come from client organisations themselves. For example, client organisations started to streamline their business processes and operations by relying on fewer suppliers with a much improved relationship, i.e. partnering with fewer suppliers to guarantee quality and low prices. Such shifting focus has created more demand on activities such as electronic trading (e-commerce), standardisation in business processes across the supply chain, and better and easier IT tools to support and coordinate the massive amount of information and activities among the supply chain's partners.

### Phase IV: Internet focus

The Internet started to provide a reliable communication infrastructure for organisations not only for the use of internal communication but also to facilitate cost-effective business improvement. The option of expanding businesses across the globe to ensure efficient and effective operations supported by appropriate and high quality international expertise is becoming more attractive to organisations. For example, a petrochemical company has its Head Office in a European capital while its specialised design offices are spread across Europe and South East Asia. The company uses the Internet infrastructure to exchange design information, manage and coordinate the design process and to communicate with its geographically dispersed clients around the clock. This process ensures a 24 hour operation without inducing high operational costs.

However, business applications over the Internet are still facing many challenges such as:

1   business process standardisation to facilitate better integration of partners' applications
2   different maturity levels of IT in organisations across the supply chain making it extremely difficult to successfully adopt advanced IT across organisations
3   the lack of effective e-procurement systems to facilitate legal aspects for trading and doing business electronically.

## 1.3  Technology push vs business pull

This section addresses some of the problems that construction and engineering industry has faced during the past decades. The lessons learnt from the above four phases is that the hype of IT (as technology) has diminished and that a new thinking has started to emerge which considers IT as an integral part of the business infrastructure. This requires organisations to effectively invest in IT to maximise their business performance and innovation. This is not a straightforward process. It might seem at first that large investment in IT could bring about large benefits to organisations. This is not always true. On the contrary, previous experience has shown that large "ad-hoc" investments in IT have caused a wide spread of isolated applications with no communication strings. An isolated investment may satisfy a particular downstream business process or function, but its overall contribution to the business may not be significant. Hence, its effective use becomes very limited, its benefit becomes transparent to management.

This problem is attributed to the business-pull and technology-push paradigms. The business-pull paradigm causes organisations to respond to the needs of the marketplace through the deployment of relevant IT, whereas the technology-push paradigm provides organisations with new technology-based business opportunities that can lead to new innovations. Under the business-pull paradigm, improvements and innovation progress are initiated by relying on strong market need alone. In an industry like construction, it is difficult to achieve innovation and performance improvement by relying on business pull alone. For example, small firms are the mainstream commodity, with 90 per cent of companies employing fewer than 10 members of staff. This does not actively promote radical change to take place at firm level.

Many firms in the industry, both at national and international levels, have successfully managed to introduce several IT applications into their organisations that have been successful in other industries. This type of technology push has been gathering momentum, and many initiatives have been led by large construction firms. In this context, the biggest challenge faced by organisations is how to integrate these technologies into their current business processes to maximise benefits to the business and to minimise risk of failure.

Both researchers and practitioners alike have recognised this fact after having encountered the problems described below.

### 1.3.1 Isolated computer applications

Since personal computers became available in the mid 1980s, their use in businesses increased and has reached a status where this technology underpins most business functions. The low cost of powerful hardware along with the availability of commercial business packages have resulted in a wide spread of isolated applications. This has led to the creation of what was known, in the 1990s, as "islands of automation", i.e. isolated computer applications which satisfy specific tasks within a department in an organisation. For example, packages such as design and analysis, planning, computer aided design (CAD), estimating, etc., are now widely used in industry to automate specific tasks which are normally carried out by individual professionals. These applications more often lack the capability to communicate and exchange data between them. When managers started to realise that they could gain further benefits from integrating such packages they were confronted with the lack of inter-operability of the different packages. This has created a tremendous burden on organisations to introduce future improvements to the tasks themselves or to the business as a whole. Case study 1.1 illustrates this point.

### Case study 1.1: Incompatibility of software packages

At an organisation which is involved in design and construction of new homes in the UK, a decision was taken at the structural design department to buy an analysis and design package to assist structural engineers in performing their design task. After a few months of market search, the department agreed to buy a particular software which seemed to satisfy the designers' needs and requirements. A number of licences were bought together with the necessary hardware.

Staff were sent on training courses to develop the necessary skills for effective use of this package. A few months later it was discovered that the output of this package could not be communicated to the estimating department within the same organisation which already had its own estimating computer packages. This is to say that the computer packages used in the estimating department could not read the schedules of materials produced by the design package due to the different type and format of the information produced. This forced the structural department to abandon the design package and to perform the task manually so that the right information could be produced for the estimating department.

### 1.3.2 Lack of communications

Isolated applications have resulted in a broad spread of stand-alone application packages with no or "fixed" communication links. To illustrate this issue Figure 1.2 shows an example of business functions carried out by a design department in an organisation. It shows the possible communication link between the various functions within the department, e.g. design, drafting, client briefing, storing and administration. Any attempt to improve one function without considering its relationship with (influence on) other related functions within the same department or other departments in the organisation can only be of a limited effect. In fact, this may have a negative result on the overall business, i.e. information cannot be exchanged effectively between business functions and as a result professionals cannot be constantly informed or updated on project changes, for example, and therefore consistency in project information cannot be maintained which could lead to catastrophic results.

The general awareness of this problem among professionals has led software vendors to provide solutions to ease the problem of communications between

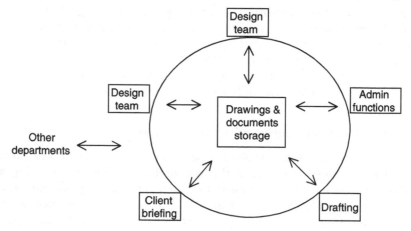

*Figure 1.2* Business functions within a department.

the existing packages. Direct interfacing of software packages has been accepted by the industry as a way of improving data transfer between applications. For example, project planning software can import costing data from a cost database or a spreadsheet either at run time or through a third party format (e.g. Dbase format). Although this approach has satisfied the need of a number of IT applications, it has highlighted (Alshawi and Hassan 1999):

1   the need for standards for data exchange that are independent of software packages;
2   the limitation of this approach in managing and controlling the exchange of project information between the various professionals.

### 1.3.3 Business processes

Technological developments and innovation have led to the introduction of advanced IT solutions to business problems including those related to transfer and exchange of project information between partners. On the other hand, business pull has generated growing awareness among organisations of the benefits and importance of process improvements and re-engineering in improving the overall business performance. The interdependence of these two issues has placed the industry in a difficult position particularly in deciding how best to bring about business improvement with minimal risk, i.e. risk of failure.

All studies in this field indicate that technology push (i.e. technological solutions) is not sufficient to improve the efficiency and effectiveness of work environments without carefully considering improvement to current business processes. Best business improvements can be achieved by considering these two issues simultaneously. This requires considerable skills and general awareness among professionals if it is to be carried out effectively (see Chapter 2). A gradual and steady movement to achieve this target reduces risk of failure and hence is more attractive to organisations. Thus, understanding business processes becomes critical and must be carried out in advance of any IT investment decision in order to improve the opportunity to explore the full potential of technology in support of the overall business strategy.

## 1.4 IS/IT investments and organisation effectiveness

Nevertheless, many construction and engineering organisations are increasingly using IT to support their business strategies (BS) (Andresen 2000). This support often extends to include a number of areas, ranging from a functional role in delivering operational needs, through a supportive role in fulfilling management and administrative needs, or an ability to deliver strategic advantage opportunities.

The "acceptable" level of IT to meet organisational needs should first be determined. This process must try to assess a range of critical issues needed to ensure a balance between the organisation readiness (mainly factors required to adapt to the proposed change) and the level and complexity of the proposed IT to ensure IT benefits are realised. This balance often includes many issues such as capital expenditure, resource availability, organisation's maturity and readiness, culture and vision, and IT training (the representation of which is shown in Figure 1.3).

Managers must therefore try to determine the exact nature, scope and influence these variables might have on success. For example, organisational culture is a variable that often consumes a considerable amount of corporate time and effort to ensure employees' attitudes and opinions are in line with the aspirations of the business. If insufficient energy is placed in this area, this may alter the balance of equilibrium, the result of which could influence the effectiveness of any IT initiative. The same can be said for all other areas, including IT training – as failure to invest corporate energy in securing appropriate training for employees can also influence overall effectiveness (Ahmad et al. 1995).

It is therefore important that the scope and purpose of any IT investment is fully justified (in business terms) and that all implementation and organisational issues are carefully assessed and evaluated before any resources are committed. This analysis should not only include provision for operational, managerial and executive needs, but also appropriate consideration of other factors – most notably the impact and influence of change, as this driver can often affect the direction and focus of future business activities.

### 1.4.1 Investment decisions

Organisations invest in IT for a variety reasons, some of which include the aspiration to improve competitiveness or to maintain market share (Tan 1996); others strive to create strategic advantage (Betts 1992); or deliver IT capability (Rockart et al. 1996). Investment therefore, not only includes the fiscal aspects, it also embraces the corporate critical time and energy (identified in Figure 1.3) needed to ensure the anticipated benefits are realised. However, the current rate and pace of change in technology and skills required to develop and manage large systems often makes it difficult to budget capital expenditure against

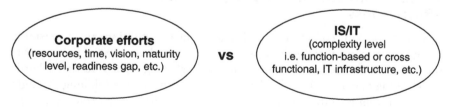

Figure 1.3 Relationship and balance between organisation energy to IT effectiveness.

current and future IT needs. As an alternative, organisations may opt to out-source their IT requirements. Whilst this may prove advantageous in some instances, this approach requires careful consideration, as it may not always be the solution for all organisations (Ward and Griffiths 1997).

Allocating resources to IT can be very difficult to justify (Heng 1996; Giaglis et al. 1999), especially where there is no strong "business case" which makes business benefits or improvements tangible and measurable. In many cases, a range of financial appraisal approaches or return on investment (ROI) models are used to evaluate IT investment, thus confining returns to financial aspects only. The use and limitations of such approaches are explained in detail in Chapter 3.

Today, however, IT is seen as an integral part of any organisation, private or public, regardless of its size and shape. Organisations cease to function when the IT infrastructure or applications that underpin their activities fail. This dependence on IT is a result of many years of IT investments in both core and supporting business functions. The quality of decisions on such investments varies from one organisation to another depending on the nature of the IT applications and the level of the organisation's maturity. Over the past years, however, IT investments have shifted from a business function focused invest-ment into strategic applications. The former was mainly concerned with small business applications that aimed to relieve a particular bottleneck, i.e. improving the efficiency of a particular business function. These small investments were either driven by:

1   demands generated at operational levels in order to satisfy particular needs (bottom-up pressure) or
2   requests issued by senior managers to meet some business requirements (top-down pressure).

In either case, there was little attention and support from senior management to their implementation. Such applications are mainly technology-focused and cost-driven ones. More recently, for organisations seeking competitive advan-tage or innovation through innovative use of IT, investment decisions are for-mally planned and aligned with business strategies. IT applications are likely to underpin organisations' core functions and are normally business focused and of high impact on business performance.

The decision-making process which leads to IT investment can have a signifi-cant impact on the success of IT applications in achieving intended objectives. Case study 1.2 demonstrates that an ad-hoc approach to IT investment could lead to an ineffective use of resources and thus to "failure" and expensive activities.

### 1.4.2 Case study 1.2: ill-informed decisions on IT investments

In 1995, a multidisciplinary contracting organisation, with a turnover of £300 million, has three divisions, the largest of which is the construction division. A software selection exercise was commissioned by the Group Marketing Director with a view to procuring a computer system to serve the various marketing managers in the construction division. It was planned that the system chosen would eventually be used across the group and would be networked around a central database. After the implementation of the selected system, it was found that it was not as powerful or flexible in use as originally envisaged. Furthermore, the senior technician who was recruited to manage the technical maintenance of the system was made redundant a few months after it was installed. As no one was trained to cover the essential work he had been carrying out, the system fell into disrepair. Furthermore, the software house which supplied the system appeared wholly incapable of providing effective support to the software package.

A newly appointed Group Marketing/Public Relations Director requested further database facilities which, on the face of it, should have been possible within the acquired system. After several weeks of deliberation, the software house announced that what was required could not be done wholly within its system and that an unspecified third party software package would be required to carry out part of the required work. Closer examination showed in fact that it would be much more cost effective to have all the work done in a third party system than to split it two ways. This approach also avoided the split responsibility issue between two suppliers and prevented more work being commissioned from the software house, a supplier who had already been shown to be suspect at best.

As a result, the new database was eventually written in house with the assistance of an outside contractor at a total development and installation cost of around £10,000. This database effectively picked up where the marketing system left off, at the award of contract stage, and gave public relations staff the information they required to support their work.

The introduction of computing systems to the marketing department has clearly not gone smoothly, and there has been a lot of expense along the way. To get an idea as to whether the investment has been worthwhile it is helpful to consider the total cost of the system per user who was usefully served by it at the time of carrying out this study. (Some of the figures are estimates where accurate data was not available.)

- The cost of setting up the marketing application package, including the software, hardware, consultancy and network installation has been estimated by the client at £120,000.
- The initial systems analysis and software evaluation took approximately 26 man weeks, at £1,000 per man week, giving a total of £26,000.

- User training was given to 12 users at a total cost of £7,500, including the users' time.
- The total cost of the senior technician who supervised the network was £725 per week for 26 weeks: total £18,850.
- The new database cost £10,000 to develop.
- Maintenance costs, including software support, staff time and an allowance for down time, are estimated at £3,000 per year over the three years the marketing system has been in place. Total £9,000.

Therefore, the total cost of the two systems has been at least £191,350, without taking into account the costs associated with the failures of the system to do its job effectively. Only four users could be found who would admit to obtaining any benefit at all from the systems and these had been actively involved, on average, for only two out of the three years since the system was first installed. So, the total annual cost of having this system is found to be at least £23,918 per user served. This figure is about six times the level originally anticipated, due primarily to the drop off in user numbers.

## 1.5 Evolving role of IS/IT in organisations

Organisations are increasingly using IT in support of their businesses and it is widely accepted that IT is becoming a key element of any organisation. In fact many researchers like to think that the level of an organisation's reliance on IT in the twenty-first century is similar to the reliance on electricity in the previous century when it was not expected for an organisation to function without electricity. Some people go further than that by fuelling this argument by stating that it is not feasible to include electricity as a part of a business case for a new venture, so the case for IT. For example, networks, Internet, e-mail and office automation are seen as standard applications for "reasonable" size organisations. For small businesses, standalone applications such as e-mail, presentations and report writing are seen to be essential components for running any business effectively. For larger organisations, the picture is more complex where IT infrastructure plays a key role in supporting core business functions. It is no longer the case to see a large organisation, for example, managing its human resources and finances through manual or isolated procedures nor pricing tenders manually using paper-based cost sheets.

The penetration of IT into industry happens as a result of many years of heavy investment that began in the 1970s. The role of information systems and their importance in improving the performance of organisations has matured over the years. The authority and role of those responsible for the implementation and management of IT have become more powerful not only in their departments but also in their involvement in the strategic business development. Somogyi and Galliers (1987) classify this evolution into three "eras": data processing, management of information systems and the strategic information systems. Each era has its own features depending on how technology is utilised to manipulate business data and information, moving from managing the delivery and operation of "back-office" isolated systems to well-managed and integrated services to various departments. This transition is accompanied by a decentralisation of IT resources to give more freedom and flexibility to business units to select and evaluate their own systems where users become more active in the development process.

However, there is ample evidence that the era of strategic information systems (IS) did not bring competitive advantage to organisations. In spite of the early studies that indicated the need for IT investment to be carefully planned and aligned with business strategies (Earl 1989; Venkatraman 1991), a large percentage of systems have failed to achieve their intended business objectives. An empirical study carried out by Clegg et al. 1997, wherein approximately 1,400 organisations were surveyed, showed that 80–90 per cent of IT investments did not meet their performance objectives and the reasons for this were rarely purely technical in origin. Moreover, the development and implementation of information systems can create an enormous amount of disruption to normal activities of organisations, thus causing unnecessary cost which is extremely difficult to measure. This dissolution in the strategic benefits of information systems is making many organisations invest in IT not for any competitive advantage but for the reasons of bringing efficiency and effectiveness to business processes (Peppard and Ward 2004).

However, it could be claimed that there are many cases where IS/IT investment has clearly brought a competitive advantage to organisations. Ciborra (1994) suggests that successful applications are mainly due to serendipity rather than to any formal planning. In fact, most of the published cases that demonstrate gaining a competitive advantage out of IT investment are "one-off" instances (Peppard and Ward 2004). Also, there is abundant evidence to show that some IT investments are ineffective and underutilised (Legge et al. 1991; Majchrzak 1988; Page et al. 1993; Xia and lee 2005; Standish Group 2001; Lientz and Larson 2004).

In recent years, it became clearer that pure reliance on technology is unlikely to bring about innovation. The wide commercial availability of powerful business software has turned them into "standard" packages which are available cost effectively to all organisations. Davenport (1998) states that direct implementation of standard packages can limit organisational ability to innovate and gain a competitive advantage in the marketplace. The driver for competitive advantage can only come from the business and business-led factors such as business changes, innovation, process improvement and new work environments. Therefore real innovation and continuous improvement can only be achieved by constant innovation in products and business practices. This requires the essential complementary organisational vision and capability within which effective innovation can be achieved. Chapter 2 addresses the success of IS/IT from two perspectives, business and technology.

## 1.6 IS/IT measurement, success and effectiveness

The main goal for any IS/IT project is the successful and timely delivery of a system that meets its planned performance and objectives. However, IS/IT projects frequently fail (Simpson 1987; Barki and Rivard 1993). It has been reported that, on average, IS/IT systems are delivered a year behind schedule, and only 1 per cent of projects finish on time and to budget (Stockman and Norris 1991). According to a 1995 Standish Group research, in the USA alone, 31.1 per cent of projects are cancelled before they finish, with a cost of $81 billion. Only 52.7 per cent of projects are completed, but with 189 per cent of their original estimated cost. Out of those, only 42 per cent of the originally proposed features and functions are fulfilled. The situation is made even worse by the lack of credible measurements of success for those systems that are completed.

Measuring the success of information systems has been a puzzling question since the introduction of computers into the business environment (Hoos 1960). A great amount of effort has gone into finding an answer to this question by looking at what effects information systems have on individual "white" and "blue" collar workers (Roach 1987, 1991), management of different levels, groups of different types, and organisations of different sizes, types, and objectives. Those were looked at from different points of view – economic, financial or non-financial. Also, different levels of study were conducted across different firms, sectors and industries, entire economies or nationally, and internationally (Brynjolfsson and Yang 1996; Strassmann 1997).

Answering such a question first gained momentum among economists in the 1980s when they tried to find the economic "hard" IS/IT value to business. This effort did not prove successful. This economic research dilemma was then termed the "productivity paradox" the meaning of which was summarised by the Nobel Prize economist Robert Solow in the *New York Times Book Review* of July 1987 as: "We see computer age everywhere except in the productivity statistics". This paradox spanned over more than a decade and left a large body of research, and its effect is still seen in IS/IT research (Brynjolfsson and Hitt 1998; Brynjolfsson and Yang 1996; Strassmann 1997).

The productivity paradox drew the attention of IT/IS professionals and researchers, and prompted some voices to call for mobilisation to counter the arguments of the "IT critics". This mobilisation encouraged other researchers to study the non-financial "soft" value and implications of IS/IT on users and organisations. DeLone and McLean (1992), in a study that gained wide prominence and debate, tried to formulate a global outcome measure of IS/IT success employing the financial "hard" and the non-financial "soft" measures. In spite of its promising approach in resolving the IS/IT measurement problem, it generated a considerable amount of debate and criticism in the IS/IT field (Garrity and Sanders 1998a and 1998b) further fuelling attempts to find a link and measure information systems against business performance and productivity.

The following subsections will touch upon the definitions of information systems/information technology "effectiveness" and "success" and outline the existing approaches that address the IS/IT measurement problem.

### 1.6.1 Definitions of IS/IT

The terms "information systems" (IS) and "information technology" (IT) are used interchangeably in the literature (Al-Ghani 1998). One can find different definitions for both terms, some of which are overlapping. Information technology can be defined in various ways. One of the most common definitions among economists is that of the US Bureau of Economics Analysis (BEA) "Office, Computing and Accounting Machinery which consists primarily of computers". According to BEA, "Information Processing Equipment (IPE)" includes communications equipment, scientific and engineering instruments, photocopiers and related equipment, as well as software and related services which are sometimes included in the IT capital (Brynjolfsson and Yang 1996).

Other definitions of IT provided by the Institute for Development Policy and Management at the University of Manchester (IDPM 2002) are computers and telecommunications or electronic means by which to accept, store, process, output and transmit information. Other definitions exist but they are mainly variations of the above stated definitions, e.g. IT is a term that encompasses all forms of technology used to create, store, exchange, and use information in its various forms (TechTarget 2002).

Information systems, on the other hand, have been described as consisting of hardware, software, communication networks, data or information, people or participants, and procedures or work processes (Gasser 1986; Strassmann 1997).

Another definition of an information system indicates that it is a system in which human participants perform a business process using information, hardware, and software to capture, transmit, store, retrieve, manipulate, or/and display information for internal or external customers (Alter 1996). The main elements of this definition are:

> *Hardware* refers to the devices and other physical equipment involved in processing information, such as computers, workstations, physical networks, and data storage and transmission devices.
>
> *Software* refers to the computer programs that interpret participant inputs and control the hardware. Software includes operating systems and end user application software.
>
> *Participants* are people who do the work. Human participants in systems typically play essential roles such as entering, processing or using the information in the system.

*User* refers to the internal or external customers that use the output of the information system. Also, the term "user" is used loosely to refer to participant, internal and external customers. This is mainly because in many systems, employees play both roles. Nevertheless, both roles are not bound together and often participants and users are different

*Secondary customers* are people who receive some benefits from the information system even if they do not receive and use its outputs directly.

### 1.6.2 IS/IT success and effectiveness

The terms "IS success" and "IS effectiveness" were both used in the 1980s to reflect the performance of information systems. "IS success" is variously described as improved productivity (Bailey and Pearson 1983), changes in organisational effectiveness, utility in decision-making (Ives et al. 1983) and higher relative value or net utility of a means of inquiry (Swanson 1974). On the other hand, the term "IS effectiveness" is considered to refer to the result of comparing IS performance to its predefined objectives (Hamilton and Chervany 1981), while, "IS success" is determined by its achievement of those objectives.

The measurement of information systems was performed on the technical level using technical attributes that focus on performance characteristics such as resource utilisation for developing the system, hardware utilisation efficiency, reliability, response time and ease of terminal use (DeLone and McLean 1992; Hamilton and Chervany 1981). Other studies use different measures of success, such as the extent to which the information system is used by management (Cerullo 1980; Ginzberg 1981), or the impact of an IS on individual or organisational performance (Cerullo 1980; Hamilton and Chervany 1981).

In the 1990s, the difference between the two terms "effectiveness" and "success" blurred. IS success and IS effectiveness described the desired state of an information system mainly on the usage level, using factors such as use and user satisfaction (Seddon 1997; Garrity and Sanders 1998). Those that used the term success and those that used the term effectiveness cited each other, which indicated that both understood the terms to be synonymous (Ballantine et al. 1996; Munshi 1996; Woodroof and Kasper 1998). In some instances, IS success or IS effectiveness was used to mean the desired state of the IS function, i.e. the control and management of the IS in organisations, rather than the information system in an organisation (Myers et al. 1998).

## 1.7 IT uptake: the case of the construction industry

The construction industry has invested significantly in IT over the past decade which has led to a certain level of innovation and business improvements in this sector. Unlike other industries, the construction industry is a project-oriented one where teams of companies get together to design and construct a project and are dismantled when the project is completed. In spite of its high turnover which represents about 10 per cent of the Gross Domestic Product (GDP) of most countries (Olomolaiye et al. 1998), its main functions and processes are still relatively unchanged. The challenge to improve performance and reduce costs has prompted developments in a number of areas with IT being a major element in performance improvement. For example, Marsh and Flanagan (2000) state that developments in electronic commerce offer the potential to radically change the structure of the construction industry, and the way information is passed between organisations. Although the potential to improve performance exists, efforts are often hampered by the construction industry's structure, fragmented supply chain, and undercapitalisation. Furthermore, construction is considered an information intensive industry. In this context, the current state – in terms of project information storage, retrieval and transfer – often lacks an integrated and a coordinated approach. The industry accepts that transformation is a major challenge that needs to be addressed, the nuance of which often has a steep learning curve. For example, Caldas and Soibelman state:

> When a construction manager wants to find all available information about one construction activity, the drawings are in a CAD file, the specifications are in a text document, the cost estimates in a spreadsheet, the schedule is in a particular application format, the contracts in a text document, and price quotes in different websites, the major task will be how to index, retrieve and integrate information from these different media. Also to make it more complicated, these data may be stored in different organisations, with different data formats.
>
> (2000: 397)

New technologies have been developed to address some of the increasing needs of more effective and efficient communications and to bring together the widely dispersed project participants and multinational project teams. On one hand, there is a perception that the Internet provides an automatic solution to the fractured communication system in construction, in that they (the stakeholders) can adopt a "plug and play" attitude towards the use of the Internet. This perception naturally over-emphasises the role of technology, and ignores other important business aspects that need to be taken into consideration in making the overall communication process more effective and efficient. Another reason for the Internet's popularity in the construction industry over the past few years is driven by its ability to transfer information and project documents across the supply chain without interfering with the organisation's internal processes, thus giving the comfort of not challenging the internal "culture". However, the success of the Internet applications has led to the evolution of various business models, which started to draw upon links between the use of the Internet to perform project management, and how to improve an organisation's internal processes on their road to achieving excellence.

The next sections present a number of IT developments that have had an impact on performance, particularly, on achieving transformation in the construction industry. The role of the Internet is explored as an example, and it is shown how this has triggered improvements in project collaboration.

### 1.7.1 Web-based developments for project collaboration

This section addresses how web-based developments have created an impact on project collaboration in industry. According to the computer weekly (see <http://www.computerweekly.com/Article103156.htm>), project collaboration has been well received in the construction industry, as it has long suffered from complex supply chains involving architects, builders, designers and engineers. Therefore, a lot of web-based service providers in the construction industry have focused their attention on developing this area of competence. This section builds up the business case for web-based collaboration, and discusses the impact of this development in terms of the currently available collaborative software.

Alshawi and Ingirige (2003) address the area of web-based development, noting that communication plays a vital role in solving problems in project management. Scanlin (1998) also points out that communication consumes about 75 per cent to 90 per cent of a project manager's time, therefore, information needs to be current, and available "on demand". Biggs (1997) also lists communication as the root cause of most project failures, but highlights that the latest web-based solutions can link with e-mail or collaborative software which can reduce the incidence of people-related issues. Deng et al. (2001) point out that the extensive physical distance between project participants, sometimes extended over national boundaries, is the main cause of delays in decision-making. Wide communication problems – ranging from delays, through to distortion of the message – can result in cost and time overruns in projects. Furthermore, the dismissive nature of expenditure on making long-distance telephone calls, facsimile transmissions, etc., have made the project management community in construction look for more viable alternatives.

The Gartner Group (see <http://www3.gartner.com/init>) identified that the highest level of interaction across organisations generally occurs between the middle level managers known as "knowledge workers". The interactions between the knowledge workers who are working with the available collaboration tools, are therefore likely to generate the highest potential return on investment (ROI) for the project. The Group's representation of the interactions at various levels is shown in Figure 1.4.

Although there are limited contacts between the top executives, it is the knowledge workers who collaborate more regularly on the day-to-day running of the project. The Gartner Group found that most of the collaborative IT tools, such as e-mail and web-based tools, are widely being used by the knowledge

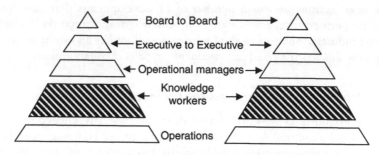

*Figure 1.4* Human interactions at various levels between two organisations.

workers. As a result of this emphasis on communication, new technologies have been developed for networking, information sharing, database management systems, etc. In contrast to the manufacturing and retail industries (where most of the new developments have taken place), the overall construction industry has shown a relatively slow up-take of web-based technologies to improve its practices (Building 2001). This situation is changing, however, as more and more firms in the construction industry are starting to realise the benefits of improving communications between project participants – which can lead to improved cost efficiency, better quality and improved competitive advantage.

It is expected that the Internet will be used to leverage even greater potential to project managers in the future. The Internet is increasingly providing a conduit for rapid information transfer, so messages not only can reach the recipients more speedily and accurately, but also are traceable to the sender (which is increasingly being used for contractual reasons). This ease of transmission can also save money for construction companies, especially when having to communicate with overseas construction entities, as the cost of providing and maintaining Internet provision is often much less than couriering documents or continual international direct dial (IDD) telephone usage. Internet provision also provides other benefits, especially when high data volumes are expected, as it does not have restrictions on locations, time or different computer operating systems.

The rapid evolution of communication technologies is making distributed projects increasingly more viable (Ly 1997). Project participants are often widely dispersed, yet they can be coordinated by sophisticated tools. The increasing availability and the usage of the Internet by small to very large-scale construction organisations has enabled project management to be performed over the web.

There are many other technological advances which have enabled companies to adopt project management over the web. Virtual meetings and tele/audio-conferencing are developments that have been adopted by some of the project management software. Ellis (see <http://www.newgrange.org>) cites that the technologies for tele-conferencing in the past often fell short of being fully effective as a result of the very high initial costs and transmission charges. However, with today's availability of the Internet, higher central processing unit (CPU) speeds of computers, and faster modem speeds (broadband), it is possible to connect one boardroom to another seamlessly, anywhere in the world.

All parts of the project can become one community using tele-conferencing and other collaborative tools. Tele- and audio-conferencing can bring the widely dispersed project team together. However, the Internet's ability as a technology to bring the diverse participants of a construction project into an effective collaborative environment is often constrained, i.e. the Internet alone cannot create an environment of interoperability among the various participants (Alshawi 2000). This is mainly attributed to the type and format of the exchanged data/documents, as well as to the different hardware and software systems in organisations.

### 1.7.2 Data exchange in collaborative environments

The type of information that can be exchanged between project applications can be categorised into two primary groups; the document group (categorised as electronic document management or EDM, and the element/object (categorised as the project model). Information transferred at a document level is normally considered as an "image", which can only be viewed, shared and annotated by users. Information transferred at the element or object level is normally detailed, and can have an impact on the outcome of the application. For example, information about a beam or a column can be exchanged between two design packages. Any changes to this type of information will automatically have an impact on other information, which exists in the other application.

Exchanging information at the element or object level is far more difficult to manage compared to exchanging information at the document level. The former needs compatible hardware and software, where information needs to be read and freely transferred between the applications. It also often requires common standards, which enable information to be transferred freely between applications. Example of such standards is the industry foundation classes (IFC). The document level uses standards such as the Internet to transfer documents as attachments to a web-based application. Most existing web-based project management software is now based on exchanging and sharing documents (and data). Information is exchanged as documents, which can be stored in a single database, that users can view, track and manipulate as and when required.

Currently, there are many commercially available software packages to cater for different types of document-based exchange, all coming under the umbrella of web-based developments. This type of software encompasses the application needs of the different stages of the project life cycle, i.e. the tender stage (where tender documents are exchanged between clients, contractors, subcontractors, etc.) and the design and construction stages (where drawings and other documents are exchanged between project partners, and where buying and selling of building materials can take place over the Internet).

Some believe that information technology has made communication worse rather than better. This is because incompatible systems used by individual disciplines can create artificial barriers that did not exist before. Within the separate domains of the design, construction, and operations of buildings, computer tools have been applied to automating specific tasks rather than addressing the overall building process. For example, the design team may use computer aided design (CAD) to produce the drawings and other documentation required for bidding and construction, but these digital products are not necessarily useful for the tasks performed by the contractor (vis-à-vis costing and scheduling). To make matters worse, the propensity and diversity of different file formats used by most CAD packages can still not always be read by some project management software. As a result, much exchange of information is still reduced to paper, even when the work was originally produced on a computer. The floor area of a building, for example, which the architect's CAD package can easily calculate, will often be recalculated and re-entered by an estimator simply because the architect's CAD package cannot easily exchange data with the estimating package. Similarly, the contractor's software cannot often easily exchange data with tools for energy analysis or facilities management. This means that so much more additional time and effort is needlessly expended in re-entering data into new systems. Furthermore, whenever information hand-ons occur, the opportunity for delay and error is much greater. The building process would be much better served if the entire chain of information – from design to construction to operations – could remain in one seamless digital format.

Over the past few years, the construction industry has witnessed the emergence of a number of powerful web-enabled software tools which can monitor, control, manipulate and store project information (see Figure 1.5). This also has an added benefit of making project information available to all participants. Many of these software tools cover a wide range of facilities and functionalities, the attributes of which have made the management process of construction projects more cost effective and efficient. Some software offers more comprehensive solutions for the entire life cycle of the project than others. In this context, abridged examples of commercially available software are presented in Table 1.1, along with their website addresses, so that their capabilities for collaboration through a project can be reviewed.

There have been many success stories where web-enabled project management has been implemented in the construction industry. This section reviews a few documented case studies (the contributors' identities have been concealed for confidentiality reasons) (Alshawi and Ingirige 2003).

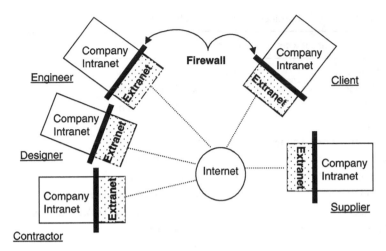

*Figure 1.5* Collaborative environments (Extranets).


### 1.7.3 Case study 1.3: collaborative environments

ALCOA – source <http://www.newgrange.org/tools_for_the_virtual_project_ma.htm>


### Background

The main objective of the project was to make computer connectivity available to every ALCOA location worldwide so that the team could collaborate over the web. The project team consisted of ALCOA (a major aluminium company in the USA), Microsoft, Hewlett Packard (HP) and the suppliers and distributors of ALCOA. The project team decided to pursue a web solution, although not explicitly stated in their terms of reference. The approach meets the monthly reporting requirements and also attempts to delight users with point and click navigation capabilities. In some respects it was an over-delivery of client expectations with the intent being to reduce monthly reporting costs and simplify a repetitive process.


### Benefits

1   Paper reports are eliminated as all the information is sent in electronic form via the web.
2   The team was able to leverage existing firewall solutions to prevent unauthorised access to the web server.
3   Repetitive routine processes were automated.
4   Data was shared among all support entities.

*Table 1.1* Web-enabled project management software

| Project management software | Website address |
| --- | --- |
| ProjectsOnline, | *Http://www.buildonline.com* |
| TenderOnline | *Http://www.buildonline.com* |
| TradeOnline | *Http://www.buildonline.com* |
| SuppliersOnline | *Http://www.buildonline.com* |
| Project Information Channel | *Http://www.thebiw.com* |
| TeamPlay | *Http://www.primavera.com* |
| PrimeContract | *Http://www.primavera.com* |
| Open Plan | *Http://www.welcome.com/products/opp* |
| Project.Net | *http://project.net/scripts/SalSAPI.dll/website/index.jsp* |
| 4Projects | *Http://www.4projects.com* |
| Architec.net | *Http://www.architec.net* |
| Integration | *Http://www.integration.arup.com* |
| iProNET | *http://pronet.wsatkins.co.uk* |
| Viecon | *Http://www.viecon.com* |
| ProjectCenter | *Http://corporate.bricsnet.com/about/solutions/design/ projectcenter/* |
| ProjectPoint | *Http://buzzaw.com* |
| Cadweb.net | *Http://cadweb.co.uk* |
| ProjectLink | *Http://www.opentext.com/affinity/listing/causeway.html* |
| Business Collaborator | *Http://www.enviros.com/bc/* |
| eProject" Express | *Http://www.eproject.com* |
| ActiveProject | *Http://www.frametech.com* |
| DOCS Fusion | *Http://www.hummingbird.com* |
| iScraper | *Http://store.yahoo.com/riba-links/iscraper.html* |
| ProjectTalk | *Http://projecttalk.com* |
| LiveLink | *http://www.opentext.net* |
| ProjectVillage | *Http://www.projectvillage.com* |
| WebWorks" and "eReview | *http://www.web4engineers.com/webapps/index.asp* |

Source http://www.nceplus.co.uk/proj_collaboration

5    An integrated communication process promotes more accurate information transfer.

## Problems

1    The firewall protection needs to be upgraded from time to time.
2    There is a need for a project manager with IT infrastructure knowledge.

3    The web server must be treated as a full-scale IT project, entailing:

- appropriate funding
- appropriate staffing
- utilisation of project management knowledge.

4    The corporate culture must be understood in order to prevent clashes.

### 1.7.4 Case study 1.4: collaborative environments

INMANCO – source <http://www.thebiw.com> <http://www.cite.org.uk>

#### Background

This case is concerned with an electronic document management system that is capable of facilitating, reviewing and updating project drawings and specifications over the World Wide Web using the Project Information Channel (PIC) of the Building Information Warehouse. The cost savings mainly arise from not having to print drawings and specifications from time to time and the capability to track changes electronically.

#### Benefits

The case study shows a £58,130 saving on a £5 million, 30-week retail construction project, involving an international firm of management and construction consultants (INMANCO). The detail breakdown of the cost saving is as follows:

| | |
|---|---|
| Printing costs for project drawings | £46,112 |
| Postage costs for project drawings | £1,584 |
| Copying costs and project specifications | £10,215 |
| Postage costs and project specifications | £219 |
| TOTAL | £58,130 |

These are only the identified direct cost savings. Potentially much greater savings were achieved through reduction in mistakes and re-works and by avoiding unnecessary project delays. In the same study, the company estimated these additional indirect savings to be in the region of £300,000, which is 6 per cent of the overall project costs. They were able to achieve this saving through the following means:

1    Delays were avoided because team members did not have to wait for the arrival of updated drawings and comments, and requests for information were immediately delivered to the headlines page of the relevant team members.

2    Visits to site and travelling time to meetings were reduced because the most up-to-date progress photographs were always available for viewing on the system.

3    Mistakes were avoided because all drawings and documents were always up to date and instantly available. There was no longer any risk that team members were acting on information that was out of date or incomplete.

4    Money spent on disputes was minimised as the system creates a full audit trail containing all the minutes of meetings.

### 1.7.5 Case study 1.5: collaborative environments

CATHQUARTER – source <http://www.buildonline.com>

#### Background

The case study is concerned with the use of a project collaborative tool, Projects-Online, which is web-enabled and has the capability to enhance cooperation and coordination of team members across several countries to achieve strict deadlines of projects. The Cathedral Quarter project involved collaboration between a Dublin based architect and developer, and a Northern Ireland structural engineer, services engineer and quantity surveyor.

As the architect firm issued drawings they were uploaded onto the project website, members of the project team in Belfast were notified automatically and could retrieve these documents immediately, regardless of their location. Comments could then be posted online, thus reducing the turnaround time on documents and drawings from days to minutes. ProjectsOnline managed the team members' access rights, ensuring that members had access only to the data relevant to their roles so that they could not make unauthorised alterations. New information posted on the website wass alerted to all team members. Any member could access or submit drawings, documents etc., and view all project information on one secure location.

#### Benefits

1    There was a substantial increase in the speed of communication, resulting in shorter lead-time on tasks.

2    The increased accuracy of communications reduced errors and rework costs.

3    Travel costs were dramaticall reduced.

4    Costs were reduced in hard copy production, distribution and storage.

### 1.7.6 Case study 1.6: collaborative environments

TOTINS – source Deng et al. 2001

#### Background

The objective of the Total Information Transfer System (TOTINS) was to help information transfer more effectively during the construction process, between Head Office and its overseas construction sites. TOTINS uses Linux (Unix based) and MS Windows 95 (PC based) for setting up the information transfer system. Logging to the remote host is achieved through telnet and transfer of files through File Transfer Protocol (FTP). Telnet protocol allows an Internet user to log into a remote host from his/her local host computer. Due to its direct connection to the remote system, the system can provide a cheap and efficient method of getting information when compared to facsimiles, phone or snail mail. FTP is a way of sending files across the Internet. This function enables file transfer from one computer to another irrespective of their operating systems. Security of the server information is provided through password access. E-mail and Internet chat is enhanced with on-screen images; pictures or drawings can simulate telephone conversations and meetings. TOTINS was applied for project monitoring in a small-scale residential project in China.

The main reason for adopting TOTINS was the extensive geographic separation of the Head Office and the site, which resulted in a very high cost being incurred for the existing project information system. Therefore an Internet based project information system was devised for data retrieval and processing by intelligent HTML and Java programming. The project information system focuses on information generation (data input at site), information transfer (web-based communication) and information retrieval (intelligent graphical view on the web). The data is submitted through a standardised form and it is saved in a text format in the Head Office web server. The Java applets specifically programmed to handle the data in text format refreshes the parameters in the web server. Performance at the project site can be compared with the schedule / estimated performance. The schedule and the actual progress is automatically plotted graphically to get a clear picture of the site progress.

#### Benefits

1   Improved efficiency was brought about by speedy and accurate transfer of information between Head Office and sites.
    Better management and decision-making were facilitated.
    There were savings on communication through the Internet as against traditional methods such as IDD phone calls and courier services.

### 1.7.7 Case study 1.7: collaborative environments

"3COM" – source Building Centre Trust 1999

#### Background

The project involves construction of office space/laboratory space for 3COM, which is an American computer networking company with offices in 45 countries around the world. Design began in July 1996 and the buildings were completed in February 1998. The client emphas was on using a partnering approach within the team to execute the project. The case explains how collaboration was achieved among the team members through electronic means via e-mail. Drawings were issued electronically as e-mail attachments to each member of the team and requests for approvals and confirmations were also accepted via the same medium. 3COM provided the network file server and group working software (lotus notes) and backup facilities.

Communication was achieved via ordinary telecommunication links with the central file server. Design consultants were only responsible for issuing an electronic drawing file to the file server and for issuing a single printed copy to the other members of the project team. The contractor was responsible for duplicating the drawings at site, which were distributed to the site team and subcontractors.

#### Benefits

1    The speed of communicating drawings increased (sending drawings by post or courier comparing unfavourably with the e-mail attachment, which takes a few minutes to download).
2    Due to the increased speed the team was able to agree an additional one million pounds worth of work without a time overrun on the project (about six weeks of time overrun was saved).
3    Traditional monthly site meetings were converted to "information required meetings" so that not all team members need participate in them. The contractor produced an electronic report and distributed it to the team members.
4    Reduction of delivery and copying costs resulted from the contractor being more focused on the precise requirements for drawing copies and in an overall saving for the project (approximately £25,000).
5    Reduction in storage space for paper work resulted of more and more electronic storage.

*Problems*

1  There were costs involved in overcoming incompatibilities.
2  Team members possessed different degrees of IT sophistication. Therefore some members had to go back to earlier versions of the software to be in line with the rest of the team. The M & E contractor preferred to use the traditional drafting system, which needed to be input into a CAD based system elsewhere and transferred back to site.
3  Technical issues such as inability to deal with large file sizes and various security issues had to be overcome. Printing of drawings at site was not straightforward as it was reported that backgrounds had to be assembled and some layers were lost in the file transfer.
4  Although the system worked well with the principal members of the team, some subcontractors found it difficult to match the technology.

# Chapter 2

# Organisational elements for IS/IT success

## 2.1 Business dynamics and technology

In the light of increasing competition, many organisations have started to use IT not only for performance improvement and cost reduction, but also to open up new markets and/or gain an advantage over their competitors. With a better understanding of IS/IT at the executive level, IS/IT strategies are being aligned with business strategies with the aim of deploying advanced systems in support of innovative business processes that best achieve the business objectives. This alignment process requires a careful and balanced approach between the level and complexity of the enabling technology and the required level (expected) process change within the organisation. Achieving this balance is a difficult process which requires highly skilled professionals who fully appreciate the strategic needs of the business and the benefits and functionalities that advances in IT can bring about to achieve the business strategy.

### 2.1.1 Relation between business process and technology: the five-layer diagram

To explain the integrative nature of business and IS/IT and to highlight the key organisational elements which underpin this concept, a five-layer model was developed (see Figure 2.1). This conceptual model clearly demonstrates the relationship between the dynamic nature of business and the supporting IT infrastructure (see Chapter 7 for details).

*Layer 1: business environment*

In a project-oriented industry like construction and engineering, where projects are at their core business, business processes are highly dynamic and could vary from one project to other even within the same organisation (see Case study 2.1). Roles and responsibilities of project partners are defined by the type of procurement method that legislates the relation between members of the supply chain. For example, the processes of designing and constructing a project in a

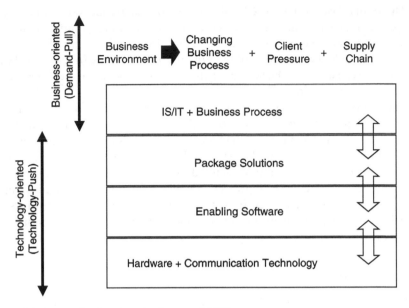

*Figure 2.1* Integration between business and IS/IT.

"partnership" agreement are different to those outlined in a traditional contract. This variation in business processes, within and among partner organisations, will not only affect the internal performance of partners but also affect the efficiency of the communication process between the partners. This complexity reflects the nature of the construction and engineering industry which makes the alignment of the IS/IT with business strategies, to achieve a competitive advantage, difficult. This is reflected by the top layer in Figure 2.1, i.e. changing business requirements.

### Layer 2: IS/IT and business processes

IS/IT systems are deployed in organisations either to support their business functions or to bring about innovation and competitive advantage. This deployment will lead to the integration of the IS/IT into the organisation's work practices which can either be accommodated by

*   introducing new business processes which are enabled by advanced IS/IT (i.e. can only be realised by the deployment of IS/IT) or
*   aligning the existing business processes with new functionalities of the IS/IT system.

(The rest of this chapter will address the weakness and strengths of these two approaches.)

The second layer in Figure 2.1 reflects the strategic business decision on the role of IS/IT in achieving the organisation's business strategy and whether the organisation has the capability to do so. Thus, this is an important layer which portrays not only the capability of the organisation to successfully deploy IS/IT but also the capability of the IS/IT to bring about improvement and competitive advantage to organisations.

The first two layers are business oriented and highly affected by the "demand pull" mechanism (Chapter 1, Section 1.3). The effective utilisation of IS/IT requires highly skilled business managers who understand both the internal needs of the organisation and the external demand on the business. In addition, they should have the capability to make decisions not only on the selection of the appropriate strategic IS/IT projects but also to assess the required changes in business processes as a result of the deployment of the new IS/IT. Such decisions are critical to achieving success and maximising the benefits of IS/IT to the organisation.

### Layer 3: package solutions

The third layer represents the software packages that support decisions made at the second layer. Once a decision is made to invest in improving a particular business process (being strategic, core, or non core) the next stage is to implement the most appropriate software solution in support of this process. This would involve decisions such as whether the software solution should be a) internally developed and implemented, b) a commercially available package or c) outsourced to a third-party. In all cases, such decisions are technology-based and require highly skilled IS/IT people with a good understanding of IS/IT implementation strategies. This layer is the main link between business decisions-makers and technology professionals, i.e. the link between the second and fourth layers. It is also prone to a "demand pull" mechanism as any selected IS/IT solution should meet the functions and specifications identified in the upper layer. It is important to mention that at this stage the "technology push" mechanism also plays a key role in that technology might identify new opportunities to further improve or even re-engineer the identified business processes through its advanced capability.

### Layer 4: enabling software

Once the software solution/package is selected – whether bespoke or commercially available – a full development and implementation plan should be established. This would include the design, development and implementation of the various components of the IS/IT solution, such as databases, application software modules and communication protocols. Decisions of this nature require experienced IT professionals, with project management skills, who can optimise the use of an organisation's resources such as the existing functionalities and capability of the organisation's IT infrastructure and its adaptation to meet the new requirements, if it can be utilised for the implementation of the new system.

The "enabling software" layer represents this process and shows that decisions of this nature are strongly affected by "technology push". Technology is normally evaluated and its impact on achieving the project's objectives is assessed. This layer creates a strong link between software development and implementation, and the hardware and communication technology.

### Layer 5: hardware and communication technology

The last layer in Figure 2.1 is the "enabling hardware" layer. This layer represents the design and implementation of the hardware infrastructure which is capable of delivering the proposed solution effectively and efficiently. This would require highly skilled professionals in hardware and communication technology such as types of network, Internet communication and hardware specifications and performance. This layer is very much subjected to "technology push" which considers the best technological solutions that are reliable, flexible and cost-effective.

This five-layer model highlights four main elements which could influence the effective selection, development and implementation of IS/IT. These elements are people and skills, business processes, IT infrastructure, and work environment. The following sections focus on each of the four elements and their impact on the successful implementation of IT in organisations.

## 2.2 Difficulties in integrating business and IS/IT: the case of construction

The government, industry and clients are all seeking to bring about change in the construction industry to improve quality, competitiveness and profitability and to increase value to clients. The implementation is undertaken through initiatives such as the Construction Task Force (Egan 1998), the Government Construction Clients Panel (GCCP), the activities of the Construction Industry Board (CIB), the Construction Clients Forum (CCF), and other CIB umbrella organisations. These initiatives are established to secure a culture of co-operation, teamwork, and continuous improvement in the performance of the industry. Where the emphasis has traditionally been on the need to manage the interface between the project and the client's organisation, it is now shifting towards the need to manage the flow of activities across the supply chain, concentrating on those activities that actually add value. Activities that facilitate the management of project information need to be identified and streamlined. This is not an easy task in construction simply because of the project-oriented nature of industry.

The construction sector operates at a wide range of sizes and types of organisations. It deals with an enormous range of projects in terms of size and complexity. The sector also draws on skills and resources of a highly varied nature starting from highly skilled designers to non-skill site workers. Though they all work in various teams to deliver one project, they are pulled together from different organisations and have different experiences and backgrounds. The design stage involves highly professional design teams constituting members such as architects, project managers, structural engineers and mechanical and electrical engineers. This is a highly information intense stage where more than 80 per cent of the project information is generated. However, much of the assembly process and materials supply parts of the construction process create much less new information but are highly vulnerable to changes in and exchange of information. The complexity of the different parties and the lack of communications between them is one of the major problems facing the construction industry.

In order to understand the cause of the problem, a simple process-focused analysis of the current situation is required. Figure 2.2 represents a hierarchical structure of a typical organisation. Each organisation has strategic or core business functions which identify its speciality and the type of activities it undertakes. Each core business function is accomplished through a number of high level functions at either a department or project level. These high level functions are of two types: business functions and supporting functions. The former has the responsibility of achieving the core business objectives by directly executing activities to achieve them. The latter is necessary to facilitate the achievement of the former, including administrative and marketing functions. Each high level function is accomplished by a large number of low level processes which are mainly concerned with low level data capturing and

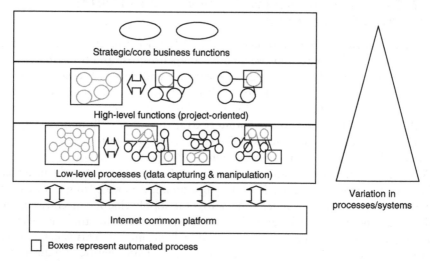

*Figure 2.2* Hierarchical structure of business processes.

manipulation. The low level processes of each high level function are normally isolated from the other low processes of a different function but they both are considered to have "client–customer" relationship. The variation in the execution of these processes increases with their location in the hierarchy (i.e. the greater the variation at operational level).

To illustrate this argument, consider a design office with a vision of providing clients with a one-stop shop for the design phase. The core business functions for such an organisation could be: architecture, structural, and mechanical and electrical design. Each of these core business functions is conducted through a representative department which runs its own specific processes and is most likely to have its own future plans and targets to achieve in isolation from other departments. Although there is a strong relationship between the three departments – as the outcome of one department is the input for the other – the level of collaboration between them or the integration of their business processes is very low. Typically, the relationship that exists between them is more of a "client–customer" type. Processes of one department are very likely to have been created based on the inherited experience and backgrounds of its employees with little or no integration with other departments. Therefore communication between the various departments is normally limited to the exchange of "formal" documents.

The picture becomes more complicated when IS/IT is considered. In the construction and engineering industry, investment in IS/IT over the years, has resulted in isolated applications, in various departments, taking place at various times. Such decisions, as explained in Chapter 1, are normally taken to automate a certain function in a department with the aim of relieving an operational bottleneck. Such a scenario leaves an organisation with different IS/IT systems which are incompatible and extremely difficult to communicate within one department let alone throughout the organisation. However, with advances in the Internet, organisations have started to use the Internet infrastructure as a communication platform between the various IT applications. Although this type of communication is beneficial to organisations in improving the speed of communication, it adds very little value, if any, to improving business performance (Alshawi 2000). For example a "building drawing" sent from the architecture department to the structural engineering department via an e-mail attachment will add no value to how designers, in both departments, can work together, e.g. will add no value to finding a solution to problems like the coordination between architecture and structure drawings.

In a construction project that involves a large number of partners with different core business functions and IS/IT applications, improvement to the flow of the project information across the supply chain becomes extremely difficult. In addition, as partners work together for a limited period of time, i.e. over the project duration, long term business improvements across the supply chain (beyond the exchange of documents over the Internet) through IS/IT are equally difficult.

### 2.2.1 Case study 2.1: lack of standardisation in business processes

The following case study demonstrates the concepts discussed in the previous sections. The scenario reflects a real situation which has taken more than six months to unfold. It clearly demonstrates the lack of processes standardisation even within the same organisations, the "technology push" mechanism which is adopted by many organisations in construction, and the influential role of people on the success of IT applications.

A large contracting organisation requested a post-contract, cost-control software package to be installed in all its UK based construction sites, as part of the business improvement plan which was agreed by the organisation's Board of Directors. A consultant (specialised software vender) was hired to implement the state-of-the-art hardware and software with the aim of improving site reporting facilities and communication to head-office. Initially, the client organisation requested a pilot study to be conducted on one construction site before approving the implementation of the software package to all its construction sites. The consultant analysed the post-contract, cost-control process on the selected construction site and found that this process had no structure and thus any attempt to use IT to improve the process would be limited and ineffective. With the assistance of the project manager, who was very open and collaborative with the consultant, the process was re-structured and the software package was implemented (after training two site staff), resulting in clear improvements in terms of providing Head Office with accurate and timely project information.

The consultant was requested to carry out another pilot study at another selected construction site of the same organisation. After facing a difficult time analysing the same process, due to the non-collaborative nature of this site's project manager who did not give the consultant free access to the site processes, the consultant had no alternative but to apply the same restructured processes which were developed at the first construction site, as both sites reports to the same Head Office. After implementation, the result was unsatisfactory where project information failed to meet the initial objectives, i.e. of providing Head Office with accurate and timely project information. After a few months of negotiation with the Head Office to give the consultant the authority to have access to the second construction site (because of the resistance of its project manager) to find out what had led to this problem, it was discovered that the post-contract, cost-control processes were carried out differently from those at the first site, i.e. they were mainly carried out according to the experience of the second project manager. Therefore, it became clear to both client and consultant that the structured process developed at the first site could not be copied to the second site.

This problem was then recognised by the organisation's Board of Directors which had commissioned the consultant to examine all the organisation's 18 construction sites. Results had shown that the post-contract, cost-control processes were carried out differently on different sites, and that the information sent to Head Office was different in type and format depending on the project manager's experience. This meant that Head Office had to treat information received from each site independently.

This case study clearly demonstrates that the absence of "standard" practices can be an obstacle to the effective implementation of IS/IT, i.e. IS/IT alone cannot be used to improve internal business performance nor can it create competitive advantage without a structured business culture first having been introduced.

## 2.3  Building IT capability

The previous section clearly demonstrates that "technology push" alone, even though to some extent still dominant in many industries like construction, will not lead to competitive advantage. Although the implementation of a few advanced IT applications might bring about "first comer" advantage to an organisation, this will not last very long as it can be easily copied by competitors. It is the innovation in process improvement and management, along with IT as an enabler, that is the only mechanism to ensure sustainable competitive advantage. This requires an organisational state of readiness which will give it the capability to positively respond to innovation in business improvement and advancement in IT.

Organisational capability is defined as the ability to initiate, absorb, develop

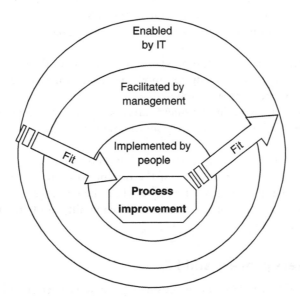

*Figure 2.3* The four organisational elements for IT success.

and implement new improvement ideas in support of an organisation's objectives. Kangas (1999) and Moingeon et al. (1998) refer to organisational capability
as the strategic applications of competencies. The development and deployment of specific organisational competencies is the process of building the
organisational capability. In the context of IS/IT and organisational capability
to ensure sustainable competitive advantage, competencies refer to many factors such as highly and flexible skills, awareness of change, flexible management
structure, process improvement schemes, clear business goals and advanced and
flexible IT infrastructure.

The competencies that an organisation needs to develop in order to acquire
the capability to strategically benefit from IS/IT fall under four main elements:
people, process, work environment and IT infrastructure. These elements are
highly interrelated, i.e. developing competencies in one element must be accompanied by improvement in the others. Figure 2.3 shows the relationship between
the four elements. Process improvement is shown as the core competency that an
organisation needs to develop to achieve the sought IS/IT capability. This element needs people with the necessary skills and power to implement process
improvements. That, of course, cannot be undertaken without the management's consent and the creation of an environment that facilitates the proposed
change through activities such as motivation, empowerment and management
of change. A high level of integration between the three elements can be enabled
by flexible and advanced IT infrastructure.

The first two elements (people and process) are the key to change and
improvements while the other two elements (IT infrastructure and work environment) are enablers without which the first two elements cannot be materialised.
Through good and effective management and advanced IT infrastructure an
attractive and innovative work environment can be created, where

- people are motivated, empowered and made aware of the expected change:
  therefore they will be ready to innovative, absorb new ideas and develop
  and implement them effectively;
- business goals and improvement targets are clearly communicated to
  employees with strong support from top management.

The time required for an organisation to build up an IS/IT capability is highly
dependent on the level of maturity of the organisation in each of these elements. The following sections will further explain the relationship between these
four elements in terms of achieving IS/IT capability and innovative work
environments.

## 2.4 Business process and IT

Davenport and Short (1990) define business process as a set of logically related
tasks performed to achieve a defined business outcome while Hinterhuber

(1995) defines business process as a set of integrated and co-ordinated activities required for producing products or offering services.

A business process has a structure, inputs, outputs, customers (internal and external) and owners and is built up by integrating fragmented tasks that contribute to its operations and internal and external flows (Hammer 1990). Business processes can be classified into four groups: core (central to business operations), support, management (concerned with organising and controlling business resources), and business network processes (with scope beyond organisational boundaries) (Willcocks and Smith 1994).

As they are based on the organisation's culture and work practices, business processes can differentiate organisations and thus can contribute to the organisation's efficiency and competitiveness. For example communicating site information to Head Office in contracting organisations can significantly differ from one organisation to another. Indeed it might differ from one site to another within the same organisation, as clearly shown in Case study 2.1. The more streamlined and effective these processes are, the more efficient the organisation is.

IS/IT can help in reshaping business processes in that it can introduce a capability to facilitate the flow of information between processes which cannot possibly by handled manually. This could be either local or globally distributed processes with ensured availability of instantaneous and consistent information across the business (Tapscott and Caston 1993; Klenke 1994). The greatest advantage resulting from IS/IT can be attained by exploiting its capabilities to create new effective business processes, rather than merely automating outdated functions (Venkatraman 1991). The concept of integrating business processes and IT is increasingly being associated with business process re-engineering (BPR). BPR is a methodical process that uses information technology to radically overhaul business process and thereby attain major business goals (Alter 1990). Organisations which have adopted BPR in conjunction with automation efforts are able to gain significant benefits from investing in new IT systems (Davenport 1993).

Another approach is to reorganise business processes around the new IS/IT system where traditional cross-functional barriers including contradictory objectives and performance measures need to be removed. In this case, it is extremely important that organisations examine the efficiency of their processes and how they can match practices embedded in software packages in order to avoid a painful struggle to integrate the two. On the other hand, changing the selected IS/IT system to fit the current business processes will slow down the implementation process, introduce dangerous bugs, and will make upgrading the system an extremely difficult task (Koch et al. 1999). When an organisation customises the software to suit its needs, the total cost of implementation rises. The more the customisation, the greater the implementation costs. Companies should keep the IT systems "as is" as much as possible to reduce the costs of customisation and future maintenance and upgrade expenses.

While IS/IT may be used to automate existing business processes, the sole automation of inefficiently designed business processes often fails (Dickinson 1997). This is widely categorised as helping to "do the wrong things faster". In either approach, the required process change mainly depends on how the management of the organisation decides to deploy the IS/IT project, i.e. implementing a third-party product or developing and implementing its "own" system.

1   In the case of opting for a third-party product, the business function will be "challenged" by the functionalities of the proposed system. Therefore the implementation process will require the organisation to change its business processes from its current practice to the one proposed by the IT system. For example, a new enterprise resource planning (ERP) can be successfully implemented to re-engineered business processes which utilise the new capabilities of the ERP (Al-Mashari 2001).

2   In the case of developing a bespoke IT system, it is very likely that the new system will mimic the practices followed by the core business processes with a slight change to the supporting business functions to accommodate the new technology. For example, an estimating system can be developed to improve the estimating process by integrating the organisation's estimating and planning departments by following the currently used practices. However, the system might introduce new procedures for data collection and input to facilitate the execution of the core estimating function.

The following section examines the organisational capability required to maximise the benefits of IS/IT investment.

### 2.4.1 Relationship between maturity in process management and IS/IT

The strong relationship between business processes and IS/IT can influence the successful implementation of systems and therefore needs to be clearly understood and managed by organisations. The level by which organisations can successfully use IT to improve their business or achieve competitive advantage depends on the level of their maturity in managing business process improvement and in managing and utilising their IT infrastructure. For example, competing design offices can develop a technical capability to use CAD systems to produce attractive 3-D models for their design. However, the level of competencies underpinning this capability is likely to distinguish an organisation from another. For instant, an organisation can use this technology to speed up the production and quality of the design, while another organisation can develop an ability to effectively communicate the CAD model to clients and/or to other IS/IT applications which could lead to business improvement.

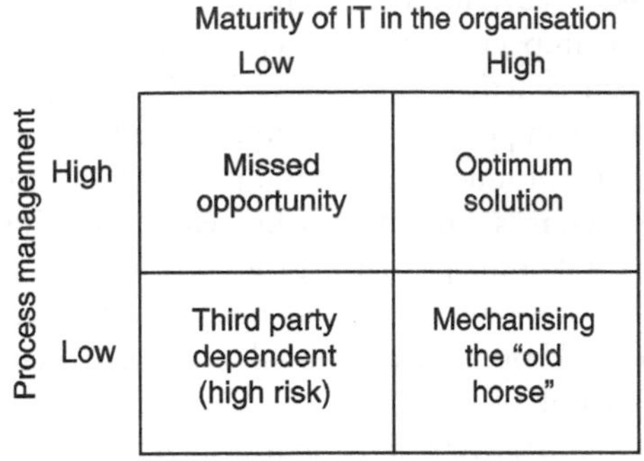

Maturity of IT in the organisation

| | | Low | High |
|---|---|---|---|
| Process management | High | Missed opportunity | Optimum solution |
| | Low | Third party dependent (high risk) | Mechanising the "old horse" |

*Figure 2.4* Relation between process and IT maturity.

Figure 2.4 portrays the nature of this relationship in four quadrants which link the level of maturity of organisations to manage process improvement with their maturity to utilise and manage IT. The top right quadrant shows that the best business benefits (i.e. "optimum solution") can be achieved from IT when the level of the organisation's maturity is high in both IT and process management. This parallel maturity allows the organisation to consider innovation and business improvement ideas in the context of advancement in IT, i.e. exploring the opportunities that advances in IT can bring about to new improvement ideas and the impact IT has on further improving the proposed ideas.

The top left quadrant shows organisations that are mature in process management but have ineffective experience in integrating IT into business processes. This type of organisation is very likely to "miss the opportunity" that IT can bring to enable better and more integrated business solutions. Although organisations in this quadrant may achieve IT successes in relieving some bottleneck business problems, they will not be able to sustain business improvements and/or competitive advantage. The lower right quadrant presents organisations with a high maturity level in IT but which have little experience in process management. This type of organisation normally invests in IT to "mechanise" the current business processes as they stand, i.e. irrespective of their efficiency. Such investments lead to management's frustration with and dissolution of the IT investments due to the inability to achieve their initial "expectations" in business improvement.

The lower left quadrant represents organisations that have little experience in both the implementation of IT and process management. Such organisations normally lack the capability to make appropriate decisions regarding business improvements and are very likely to seek external advice from a third-party (IT or management consultants). Although the latter could bring best practices to an organisation in this quadrant, successful implementation will remain difficult to achieve due to the missing internal capability.

A survey conducted by the London Business School and McKinsey in France,

Germany, UK and USA, covering 100 manufacturing companies shows that IT investments have little impact on productivity unless they are accompanied by first-class management practices (McKinsey 2005). Companies were rated on how well they used three important management practices: lean manufacturing, which cuts waste in the production process; performance management, which sets clear goals and rewards employees who reach them; and talent management, which attracts and develops high-calibre people. The companies that had the highest marks in these areas became more productive, with or without higher spending on IT. Those that combined good management practices with IT investments did best of all.

In a similar quadrant arrangement to that of Figure 2.4, the study shows that + 20 per cent of the companies surveyed falling in the top right quadrant had an increase in productivity. While 0 per cent of the companies surveyed falling into the lower left quadrant had improvement in productivity.

### 2.4.2 The relevance of the time element

Organisations can only utilise IT investments to achieve an innovative and sustainable business improvement and competitive advantage, if they create a work environment through their processes and people. Business innovation and improvement is initiated and led from within organisations by people and cannot be imported from another organisation. This can only be realised when employees are fully aware of, understand and utilise IT for the benefits of their business processes. This requires a state of readiness where people will not only have the capability to change but also can effectively utilise the available competencies to achieve strategic benefits, i.e. creating new core business functions enabled by IS/IT. This state of readiness cannot be reached without high investment in IT structure.

This state of readiness is achieved in organisations over many years of heavy investment in IT where skills and competencies from both users and management of the IT department would have evolved to reach the required state of readiness. Figure 2.5 shows the relation between IT investments/assets and the relevant business benefits and skills as seen by a petrochemical organisation. The figure reflects this organisation's experience which clearly indicates that significant business benefits can only be realised when significant investment in the IT infrastructure is in place. This investment in advance IT would lead to the development of skills and a state of readiness which would enable the organisation to innovatively utilise its IT infrastructure to improve business performance and gain competitive advantage. (See Chapter 5 for more details.)

### 2.4.3 Case study 2.2: Extranets

Organisations are using e-business tools to improve the collaboration with their partners across the supply chain. These commercially available tools have the

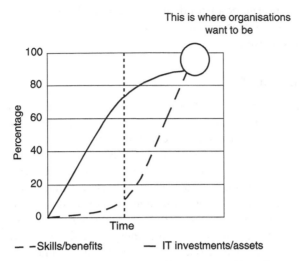

*Figure 2.5* Relation between IT investments and organisational benefits.

capacity to facilitate on-line transactions and create enhanced business efficien-
cies. Evidence of successful e-business implementation can be seen with the use
of Internet-based Collaborative Environments or "Extranets". These can be
used to span the whole supply chain (to manage, control and improve the
communication process), and can facilitate the overall operation and flow of
project information. Particular exemplars embrace remits associated with project management, tendering, supplier selection and purchasing. Such systems
have successfully captured the current supply chain communication practices
and project documents using "document management" databases provided
through an Application Service Provider (ASP), and delivered through the
Internet.

However, whilst these systems are able to deliver improved communication
across the whole supply chain, they fail to provide or help procure "value", as
they are not fully able to penetrate each organisation's value-driven chain/
processes and infrastructure (the term "value-chain" can be defined as "the core
activities an organisation performs to create and distribute its goods/services to
other members of the supply chain"). This limitation therefore creates one
major fundamental drawback, as information interchange between supply
chain partners can never really be fully exchanged at a non-document level (e.g.
accessing specification of a certain project's elements as and when required by a
partner) and thereby reduces overall flexibility. Enabling that to happen requires
developing the necessary processes, within and among partners, to allow
controlled and secured passage to the project information which all partners
have to agree upon. Consequently, various project information is unable to
be accessed and shared unless attached as documents to the communication
messages. This limitation means that organisations are unable to integrate

their suppliers' information with their upstream applications (within the existing collaborative Extranet), and therefore cannot capitalise on any benefits associated with sharing.

## 2.5 People and IT

Organisational change, by implication, involves change at group and/or individual level. Ultimately, an organisation will not be able to adapt itself successfully to a new situation, unless it can influence the behaviour of its members. Wherever the need for change is recognised within an organisation and wherever the precise nature of such change is formulated, eventually it will be up to the "people managers" to create the right environment for individual change. This means that "people managers" have a key role to play in the management of change in organisations. In this situation, organisations need to find out how to release the creative energy, intelligence and initiative of people at every level. Those that do so are far more likely to survive and prosper. They need to learn how to overcome the stultifying and disempowering effects of bureaucracy. They need to make organisations really exciting and rewarding places where people feel valued and there is opportunity to develop and grow.

Regardless of other factors, an organisation's ability to cope with its environment will not be improved unless employees alter their relationships with one another and with their jobs. Thus, any business improvement initiative is basically trying to get employees to adopt new patterns of behaviour. Consequently, in order for business improvement to be successful, new behavioural patterns consistent with that of the business improvement initiative must be developed. If this does not occur, business performance will not improve.

IT systems are heavily influenced by the social and cultural aspects of organisations. In fact, this is the biggest challenge facing organisations in the successful implementation of IT. Effective change to organisational culture and structure is considered to be an essential ingredient to IT. There is a general agreement in the literature regarding the importance of the role of people and organisational culture in delivering successful IT systems. Towers (1996) explains that managing change and people together is a major contributing factor to the success of process and IT related organisational change. Cooper and Markus (1995) state that the inadequate treatment of the human aspect is the major cause of re-engineering failure. Kennedy (1994) highlights that the elements of human change management are "the most difficult challenge" where employees feel threatened by the new processes and IT systems which can force them to change their work style. Arendt et al. (1995) show that the human factor is a major dimension that process and IT related improvements should focus on.

## 2.5.1 The role of IS/IT Department Head

The role of the IS/IT Head is important for the success of IS/IT planning, development, implementation and operation in the organisation (Weill et al. 1994; CSC 1996). However, changes in both the technology and business challenges that face organisations continue to change the role of the IS/IT executive. The Computer Sciences Corporation (CSC 1996) has suggested six IS/IT leadership roles which are required to execute IS/IT department agenda.

1  *Chief architect* designs future opportunities for the business. The primary work of the Head is to design and implement the IS/IT infrastructure so that it will expand the range of future opportunities for the business.
2  *Change leader* manages resources to achieve optimal implementation of IT. This includes providing new IS/IT tools and putting in place teams of people who can redesign roles, jobs and workflow.
3  *Product developer* helps define the organisation's place in the emerging digital economy. For example, a Head might recognise the potential for performing key business processes over an electronic linkage such as the Internet. The Head must "sell" the idea to business partners and the Board of Directors.
4  *Technology provocateur* embeds IT into the business strategy. The Head works with senior business executives to bring IS/IT and its marketplace realities to bear. The Head should be a senior business executive who understands both the business and IS/IT at a deep enough level to integrate the two perspectives in discussions about the future course of the business.
5  *Coach* teaches people to acquire the skill sets they will need for the future with two basic responsibilities: teaching people how to learn to become self-sufficient, and providing business teams with staff able to do the IS/IT-related work.
6  *Chief operating strategist* invents the future with senior management. The Head is the top IS/IT executive who is focused on the future agenda of the IS/IT organisation, i.e. strategies related to helping the business design the future, and then delivering it.

### Factors affecting the Head of IS/IT

The performance of the Head of the IS/IT department is affected by a number of factors.

1  The number of years the IS/IT department head has worked in the current organisation or the number of years s/he has been in the current position have been found to affect his/her performance. It has been argued that new, externally hired leaders will be more change-oriented than leaders who have been in the organisation for several years (Applegate and Elam 1992).

2    The extent to which the IS/IT head personally uses technology and his/her technical background also affect the performance of the IS/IT department leader. It has also been argued that the Chief Information Officer (CIO) with greater personal IS/IT familiarity and use will spend more time in technologically oriented leadership roles (Weill et al. 1994).

3    The support by top management and the relationship between the IS/IT department head and the CEO have an impact on the success of IS/IT in the organisation. It has been argued that the CIO's relationship with the CEO and other executives in the organisation who represent user groups will have an impact on IS/IT leadership roles and affect the success of IS/IT (Watson 1990). For example, the extent to which the CIO is involved in strategic planning may be dependent on the relationship of trust between the CEO and the CIO.

4    Also, the relationship between IS/IT department staff and leadership on the one hand and user personnel and their leadership on the other would affect their acceptance and use of IS/IT (Feeny et al. 1992).

5    The rank and position of the IS/IT head also affects his/her role. If there are several management layers between top management and the IS/IT head, then it can be argued that the latter's role is more managerial and less strategically oriented (Watson 1990).

### 2.5.2 Case study 2.3: people role in the success of collaborative environments

E-tendering is a way of managing the entire contract-letting process electronically. All contract documents are distributed to tenderers via a secure web-based system, avoiding the need for collating paper documents and circulating them by post. Notification is sent out to suppliers where contract documents can be downloaded from the web. Updates and queries are exchanged electronically throughout the tender phase. All the communication and exchanged documents and information are stored in a central database which is easily searchable and can be fully audited, with all activities recorded.

In a study to find out the impact of people on the success of the implementation of e-tendering in the construction industry, covering a number of large organisations, people critical success factors were found as outlined in the list below. Employees must be satisfied with the current working conditions and have the interest and motivation for self-improvement and expanding their knowledge. Employees must also be interested and motivated with IT (which has to be clearly supported by top management). Employees must also be given ample training and resources to ensure optimum productivity with the implemented system and they must be able to deal with the system as a tool that can help their performance. However, the most interesting aspect of these findings is the preference for employing a younger generation of employees which is deemed to have a better training and attitude towards IT.

*List of people-critical success factors for the successful implementation of Extranets:*

- motivation of employees
- interest in IT of employees
- work satisfaction of employees
- prior experience with Extranets
- employee attitude towards Extranets
- presence of a "champion"
- security of job – technology does not replace human processes
- internet access and type of availability
- adequacy of training
- adequacy of resources
- alignment of Extranet's implementation strategy to project team strategy
- employment of the younger generation – more interest, training and focus on IT
- proven computing technology background and capabilities using collaborative environments
- proven background in improving efficiency and productivity in work through collaborative environments.

## 2.6 Work environment

The work environment is considered to be the main enabler to the process and people elements. This element is affected by many factors of which the most important are; committed leadership, empowerment of employees, communication, process vision development, project management and process-based team formation (Al-Mashari and Zairi 2000).

### 2.6.1 Leadership

Leadership plays a vital role in directing change efforts towards success. The importance of leadership stems from its role in providing a clear vision of the future, communicating this vision, being able to involve other people in the implementation efforts and being prepared to provide sufficient commitment to the overall efforts (Hammer and Stanton 1995). Carr and Johansson (1995) also added the ability to motivate people rather than directly guide them. Additionally, Hammer and Champy (1993) describe a leader as one who is creative in thinking about change and can understand the implementation case of change and its effect on the organisation. Commitment to change and sufficient authority over all aspects of the change process are both important in dealing with the cultural and political problems in the form of organisational resistance.

### 2.6.2 Empowerment

Employee empowerment is an effective factor leading to the success of IT implementation, since it promotes self-management and collaborative team-work principles (Mumford 1995). When employees are empowered, they become more involved in deciding how work (change) should be approached and which technologies to use, and they are given the chance to participate in the change and implementation process (Arendt et al. 1995).

### 2.6.3 Communication

Communication of change and improvements is another essential tool which is perceived as very important in facilitating process and IT improvements. Carr and Johansson (1995), in their list of best practices, stress the importance of communication in successful implementation of change. However, it is also considered by organisations to be the most difficult aspect of implementation (CSC Index 1994). It is important that communication takes place throughout the change process at all levels and for all individuals, and it should occur regularly between those in charge of the change initiatives and those affected by them. It also should discuss sensitive issues, such as personnel reductions, openly and honestly.

### 2.6.4 Process vision development

Top management has to put forward an imaginative understanding of the future processes, i.e. a "vision". Process improvement and IT implementation is about redefining the company's vision, about its mission and potential customers and competitors (Clemons 1995). Process vision, as explained by Davenport (1993), describes the future state of process and therefore links business strategies with procedures and actions. He also suggests a number of tasks that make up a complete development of process vision: evaluating business strategy to anticipate future processes, benchmarking similar change efforts, conducting customer-based assessment of performance targets, and developing process attributes and their performance measures and targets.

### 2.6.5 Project management

An effective project management is considered as a critical factor in the successful implementation of IT systems. Hammer and Stanton (1995) recommend piloting the IT implementation, particularly when the project involves a large-scale organisational change. A pilot project demonstrates failures and risks involved, and provides the opportunity to make appropriate adjustments to the efforts, thus promoting success and preventing possible disasters. Risk also needs to be managed. Risk in process and IT improvement originates from incorrect or inadequate changes to processes, structure and their supporting IT systems, from making radical changes that lead to political conflicts, from investment decisions, and from introducing new hardware and software systems (Clemons 1995).

### 2.6.6 Process-based team formation

Teamwork is viewed as the most important value in implementation change (Johansson et al,1993). Davenport (1993) highlights three main functions to the change team: managing work by making group decisions and co-ordinating activities; managing relationships by promoting trust, openness, and resolving conflicts; and finally, managing exteriors such as customers, suppliers and market partners. Teamwork brings about many advantages such as facilitating interactions between functions and speeding up the redesign process (Davenport and Nohria 1994) and creating a learning environment in which team members are encouraged to share knowledge and expertise (Rohm 1992/93). Also, by empowering team members, teamwork enhances quality of work, reduces resistance to change, and allows for different perspectives to change.

### 2.6.7 Case study 2.4: work environment role in the success of collaborative environments

A collaborative environment provides a "virtual" working space to all members involved in a project where project communications and documents are shared and exchanged. The main aim of using such an environment, in project management, is to improve the decision-making process by enabling members of the project team to

1   have a quick and secure access to up-to-date project documents as and when required;
2   be able to respond easily and quickly to queries;
3   ensure that all communications are logged (stored) in a secure environment which will facilitate future "tracing back" of documents.

This means that traditional mistakes generated from someone working on an "old" document or drawing are eliminated or at the very least minimised. More crucially such an environment reduces the opportunity for mistakes leading to disputes which is the biggest cause of waste and inefficiency in the construction industry.

A large construction organisation in the UK that had been successfully using collaborative environments over most of its projects, failed to achieve the same success level with its overseas construction projects. After a careful investigation it was revealed that the organisation was working with overseas partners whose organisations are hierarchically structured with no clear employees' roles and responsibilities. People are not empowered to receive new project documents nor make relevant decisions regarding their particular profession. Any decision taken in response to a particular activity has to be endorsed by two or three levels of management and has to follow formal procedures. In addition, each of these approval steps has to be officially documented through proper channels of communication. In some cases the management which is required to approve a "professional" decision is located at a different site which means that it will take longer than "normal" to have such decisions approved.

This is a clear case where an IT solution designed to improve a business function for one culture, most likely by following best available practices, cannot be easily implemented in another culture. In this case, people felt powerless to respond to other project partners in a timely and a professional way through Extranets. This has defeated the aims for which the Extranets were designed for, as stated above. This clearly indicates that IT solutions such as Extranets are unlikely to be successfully implemented unless organisations change their work environment in line with the objectives of the IT solutions.

## 2.7  External and internal drivers for change

In recent years, companies have appreciated that business improvement and competitiveness are not based wholly on areas such as product, technology or systems. Increasingly, sustained business improvement is viewed as being generated from the "softer" dimensions of organisation: from how organisations manage their people and their processes. This shift in focus from the more traditional thinking of business improvement and success is evidenced, for example, by such developments as core competence, learning organisation, business process re-engineering and continuous improvements.

Improvements are motivated by external and/or internal drivers. External drivers are related mainly to the increased level of competition, the changes in clients' needs, technology changes, and changes in regulations. Internal drivers are mainly related to changes in both organisational strategies and structures (Al-Mashari and Zairi 2000).

### 2.7.1  External drivers

The increasing level of competition in the global market has emphasised the need for organisational innovation to cope with global standards of products and services. Therefore, co-ordinating and streamlining of the organisation's distributed functions become the main desire of many organisations seeking competitive advantage (Ovenden 1994; Bjørn-Andersen and Turner 1994). They approach process re-engineering and IT as a tool to dramatically improve their business performance and to lead them to a competitive position (Schnitt 1993; Grover et al. 1993). IT is increasingly underpinning major tasks in today's business. This is a result of its growing ability to bring new business opportunities, and facilitate the development of the new organisational forms and structures needed to meet the continuously emerging changes in business.

### 2.7.2 Internal factors

Many organisations seek strategic and structural changes that are enabled by IT and process improvement. This can be triggered off by many internal factors such as changes in terms of internal processes, methods, skill competencies, attitudes and behaviour. An example of an internal factor is the need for simplification to achieve better levels of performance, and to highlight inefficiencies which have increased as a result of increased automation in conducting business and accumulating complexities (CSC Index 1994). Also, changes in business strategy may involve some process change efforts to meet new business objectives while changes in organisational structure may necessitate a change in the underlying business processes and IT infrastructure.

# Chapter 3

# IS success measurement approaches

## 3.1 Introduction

The high percentage of IS/IT projects that have not met their intended business objectives, i.e. that have "failed", has continued to be a major concern for organisations since the 1970s. Such projects were either abandoned, significantly redirected, or even worse, they were "kept alive" in spite of their failure. The cost of funding those projects and the missed opportunities of not benefiting from their intended capabilities can represent a tremendous loss for organisations. This phenomenon has created the need for the development of evaluation methods to measure the effectiveness of IS/IT.

Previous research in the field of IS/IT success and failure considers IS/IT as a "product" which is normally evaluated, after its implementation, to measure the level of its effectiveness. The current literature classifies IS/IT evaluation methods into two categories:

1   product-based – evaluating IS/IT in terms of its technical capability and/or user satisfaction;
2   process-based – evaluating the processes that underpin the design and development of IS/IT.

It is important to make a clear distinction between the development of IS/IT systems and their implementation within organisations. The latter is concerned with the "recipient" organisation directly benefiting from the IS/IT project to improve performance or gain competitive advantage. The former is concerned with either the development of commercial systems by software vendors or with the development of bespoke applications within recipient organisations. This distinction makes it clear that process-based evaluation methods are related to the development of IS/IT projects, while product-based evaluation methods are concerned with the implementation of systems within recipient organisations. This book focuses on the latter while briefly covering the former in this chapter only.

It is worth noting that the process of planning, designing, developing and implementing IS/IT has changed over the past decade. Recipient organisations tend to lean towards the implementation of a third-party product rather than developing their own systems. This creates a different set of challenges which recipient organisations have to face. These challenges are mainly related to an organisation's ability to successfully implement IS/IT, thus creating a new demand for the development of new measurement approaches to address this issue. This chapter briefly explains the current IS/IT evaluation methods, outlining their limitation in the context of the current changes and finally explaining the various evaluation measurement approaches within the context of the IS/IT life cycle.

## 3.2 Categorisation of the current measurement methods

The current literature reveals a wide range of measurement methods, which can be divided into three categories according to the focus of the evaluation. The first category is concerned with those approaches that evaluate IS/IT as a product, while the second category refers to those approaches that evaluate the processes which underpin the development of IS/IT and the third category covers the approaches which attempt to assess the maturity of IS/IT within an organisation in terms of IS/IT planning, infrastructure, utilisation and management. Table 3.1 shows an overall summary of these approaches. The following section briefly explains each of the above three categories.

Table 3.1  Categorisation of the current IS evaluation approaches

| Approach | Type | Details/examples | |
|---|---|---|---|
| Product-based | System quality | Focuses on performance characteristics such as resource utilisation and efficiency, reliability, and response time | |
| | System use | Reflects the frequency of IS usage by users | |
| | User satisfaction | Widely used approach which is based on the level of user satisfaction | |
| Process-based | Goal centred | Measures the degree of attainment in relation to specified targets. Examples: GQM and ITIL | |
| | Comparative | Benchmarking approach | |
| | Improvement | Assesses the degree of adaptation of a process to the related changes in requirements and work environment. | |
| | Normative | a. Maturity-based<br>b. Non-maturity based | Measures performance compared to external standards. Example: CMM, ISO standards |
| Organisational maturity | General models | Examples of such models are those by Nolen; Earl; Bhabuta; and Gallier and Sutherland. | |

## 3.3 Product-based measurement approach

These measurement approaches are concerned mainly with assessing the different features of IS/IT, which include the following.

- *System quality (technical)* focuses on performance characteristics such as resource utilisation, hardware utilisation efficiency, reliability, response time, ease of terminal use (Kriebel and Raviv 1980)
- *System use* reflects the frequency of IS usage by users in a voluntary setting (Bonner 1995)
- *User satisfaction* reflects the level of user satisfaction: this is the most widely used approach (DeLone and McLean 1992).

In addition, a product-based measurement approach covers financial evaluation of IS/IT such as:

- Return-on-investment approaches (Radcliffe 1982)
- Cost-benefit analysis (King and Schrems 1978)
- Multi-objective multi-criteria methods (Chandler 1982)
- Return-on-management (Strassman 1997)
- Information economics (Farbey et al. 1992).

### 3.3.1 Limitation of product-based measures

Both the financial-based measures and the non-financial measures are still inconclusive, especially in measuring the success of IS/IT projects at the business level. These measures have been criticised for not being able to provide value to organisations. In fact, it was advocated that a major problem with IS/IT investments is the way in which they are evaluated (Strassman 1997; Willcocks and Lester 1996).

The financial measures do suffer from limitations in capturing the actual value of IS/IT in organisations. As Farbey et al. (1994: 278) explain "We argued that the problem (of IT investment evaluation) was inherently difficult because of the unpredictability of the impact and the impossibility of attributing the end result solely to the investment in IT".

The non-financial measures also suffer from many limitations. On the technical level, a technically sound IS/IT does not guarantee that it will be accepted or used by users, or that it will meet its planned objectives. On the usage level, system use is only suitable for systems that are for voluntary use which is not common in today's work environment. In addition, interfering factors, such as the individual user's work life, might affect the measurement of system use. This makes system use more suitable for controlled laboratory settings than the real world (DeLone and McLean 1992). User satisfaction has also been criticised, among other things, for not being a suitable surrogate for IS/IT success. It is not necessarily the case that satisfied individuals would lead to a successful system which meets its planned business objectives. In addition, there were no conclusive answers regarding whose satisfaction should be considered and with regard to what aspect of the IS/IT. As the respondents change, so might the satisfaction with the IS/IT (Ballantine et al. 1996; Delone and McLean 1992).

## 3.4 Process-based measurement approach

Process-based measurement approaches attempt to measure the effectiveness of the processes which underpin the development of IS. Examples of such approaches are the Capability Maturity Model (CMM) (Paulk et al. 1993), Trillium (Trillium 1996), Bootstrap (Kuvaja et al. 1994), SPICE (Software Process Improvement and Capability Determination) (El Emam et al. 1998), IT Infrastructure Library (ITIL) (Central Computer and Telecommunications Agency 1992), and Goal/Question/Metric (GQM) paradigm (Solingen and Berghout 1999).

The process measurement approaches can be further divided into four groups according to "how the evaluation process is conducted".

1   *Goal-centred measurement approach* This approach aims to assess the degree of attainment in relation to specified targets. Examples of this approach are the GQM and ITIL models.
2   *Comparative measurement approach* This is the "benchmarking" type, which involves comparing the success of two or more similar processes.
3   *Improvement measurement approach* This is concerned with assessing the degree of adaptation of a process to the related changes in a work environment. In practice, this approach is not performed separately but is combined with any of the other measurement approaches, mainly the normative approach, which is often represented in maturity based models (see Section 3.5 below).
4   *Normative measurement approach* This approach aims to compare the performance of a process to some external standard rather than to specific objectives or goals of an organisation. It is best used when standards for best performance/practice have been identified such as the Capability Maturity Model (CMM) and the ISO standards.

This group of measurement tools can be categorised into two sub-groups according to the "evaluation criteria". These sub-groups are the maturity and non-maturity based measures. Examples of maturity-based measures include:

1   *CMM* This is used in assessing the process capabilities of software organisations in achieving their project goals by placing the organisation on one of the five maturity levels.

2    *People-CMM* This aims to develop and manage the knowledge, experience and motivation of employees.
3    *Bootstrap* This European project aims to assess all processes and to examine whether the organisation implements them adequately.
4    *Trillium model* Based on the CMM and ISO 9000 series, this model is used as a benchmark to measure an organisation's processes capability and to compare it with best practices in industry.

Examples of the non-maturity based measures are:

1    *ISO standards* These are the oldest and most widely known standards for the consolidation and certification of processes in organisations.
2    *SPICE and TickIT* By adopting the ISO 9000 standards, these two models were produced for the software industry.

## 3.5 Maturity approach: general measurement models

This category of measurement approaches covers models which attempt to assess the IS/IT maturity in organisations with respect to IS/IT planning, infrastructure, utilisation, and management. They describe how an organisation can place itself within a particular maturity level of IT/IS planning. Examples of these models are:

1    *Nolan Model* This model assesses the maturity of IS in organisations in six "stages of growth"; initiation, contagion, control, integration, data administration, and maturity (Nolan 1979).

### Criticism of Nolan Model

The model faced criticism mainly because it could not prove its claims to represent reality, either as a means of describing the phases through which organisations pass when utilising IT, or to predict these changes (King and Kraemer 1984). King and Kraemer (1984) found that the empirical evidence for the stages is inconsistent and many of the model's assumptions are not practical, but they acknowledged the model's simplicity is behind its popularity with practitioners, researchers, and IS managers since its introduction. Drury (1983) found that the model stages did not map onto the organisations' status in the real world especially at the later stages, where he attributed this limitation to the simplicity of the model. But he accepted that individual organisations can use the model in assessing how effectively they are coping with the increasing importance of IS/IT. Others, in analysing critiques of the Nolan model, suggested that while evolution through the early stages of the model could be observed, the arrival of strategic systems in the 1980s introduced a new stage which changed the concept of how IS/IT evolves to maturity in practice.

2   *Earl Model* This model is concerned with the stages of organisational IS/IT planning maturity level (Earl 1989).
3   *Bhabuta Model* This model brings together elements from different aspects such as strategy formulation, information systems, and the management of information systems (Bhabuta 1988).
4   *Galliers and Sutherland Model* This model combines elements from previous models in a new structure describing the important IS elements in general, and the kinds of activities and organisational structures needed for an organisation to move through IT growth stages (Galliers and Sutherland 1991).

## 3.6 Limitation of current measurement approaches

The process-based and multi-approach measures suffer from many limitations. For example, they only cover specific types of maturity such as organisational processes for a specific industry – mainly the software industry – or they are used to assess only a specific aspect of IS/IT such as IS planning (Bhabuta 1988; Earl 1989), IT infrastructure (Weill and Broadbent 1999), IS/IT utilisation (Hamilton and Chervancy 1981; Nolan 1979), or management of IS function (Hirschheim et al. 1988; Galliers and Sutherland 1991).

Those measures, which are of product normative and process normative types, suffer from yet another limitation. Even though they collectively describe many elements that are related to technical, people, process and environmental issues that are relevant to the use, planning, development and maintenance of information systems as well as elements that describe environmental issues related to management style, structure and culture in the organisation, those measures do not provide the desperately needed assistance in predicting the outcome or level of success of an IS/IT project, prior to its implementation. The process of planning, designing, developing and implementing of IS has changed over the past decade. There is a trend towards the implementation of third-party products instead of bespoke products. The problems resulting from this change introduced different challenges particularly to recipient organisations. These are mainly related to the organisation rather than the product, i.e. the level of readiness of the organisation to successfully absorb the systems into its operations, including people and processes as well as problems related to their integration with existing IS/IT infrastructure. Furthermore, investments in IS/IT in recent years became more linked to the achievements of the organisation's business objectives, while IS/IT success was previously measured at either user or technical levels, and when it was measured on the business level, it was done in hard-financial terms only (Brynjolfsson and Yang 1996; DeLone and McLean 1992; Hitt and Brynjolfsson 1996; Farbey et al.,1994).

The measures of IS/IT have been mainly post-investment appraisals that try to assist management in reviewing the results of their decisions on IS/IT. This is to enable management to learn from previous lessons which should feed back in the decision-making process for future IS/IT implementation. However, the huge amount of IS/IT failure does not provide evidence that the lessons are being learnt.

To lessen the risk of failure of IS/IT, organisations need to be able to predict more accurately the outcome of those projects. The earlier this prediction can be achieved, the more likely it is that organisational changes needed to facilitate a successful system can be made. This calls for a pre-project/ pre-implementation measure that can help organisations to assess the implementation process prior to capital investment. In Part III, this book presents a measurement approach which adopts a balance between the product-based and process-based measurement approaches and will be combined by adopting both the normative and goal-centred approaches.

## 3.7 Success levels and measurement approaches

In order to understand the role and impact of each of the measurement approaches on IS/IT, it is important to relate these approaches to the evolving phases of IS/IT and to examine this relationship within the context of "success". An IS/IT project is considered a success if its outcomes are successfully achieved, i.e. its intended objectives. The outcomes (effects) are measured at three different levels:

1   the technical level measures the direct outcome of the project which is reflected by the quality and accuracy of data and information that it communicates;
2   the semantic level measures the outcome of the project from the users' perspective;
3   the influence level measures the degree of impact of the project on organisation's performance.

Figure 3.1 shows a framework which embraces the above three levels of success of an IS/IT project and their subsequent measurements. The framework consists of three phases: planning, development and implementation, and measurement. The technical level is considered under the IS/IT development and implementation phase. The measurement phase starts when the first outcome is received at the end of the technical level and at that point the development and implementation phase ends.

### 3.7.1 Planning and development and implementation phases

The framework starts by showing that the initiation of an IS/IT project is an "effect" intended by management, or on their behalf, in response to a scanning (driving force) of the external and internal environment either as part of a specific planning process or as a part of strategic information systems planning (SISP) initiative.

The scope of the intended effect varies according to the maturity of the organisation and could either be at the operational, managerial, or strategic level. As the organisation matures from the ad hoc stage, the scanning goes internal for implementing and tuning operational IS/IT with operational-based and management-based objectives. With further organisational maturity, scanning goes external again to achieve competitive advantage with a strategic view. This evolution process entails an evolutionary theme in the level of objectives and critical success factors (CSFs) that the planned IS/IT can achieve (see Section 3.7.3). As the maturity of the organisation increases, so does the level of IS/IT objectives and CSFs (Gottschalk and Khandelwal 2002).

To achieve the intended impact, the requirements of the system should be formulated. Those requirements are the intentions that management wants to convey to the individual user and/or participant. The outcome of this phase represents the beginning of the semantic level of communication (see Figure 3.1).

At this point, especially if the IS/IT project is not an outcome of a SISP, a need would arise to conduct a measurement study to determine organisational readiness for the implementation of the information system. This readiness measurement is to help ensure that the next steps leading to IS/IT development are not performed unless the information system has a positive recipient environment within the organisation which will increase its success of achieving the intended objectives. Following this readiness study, if the organisation is found to be at an appropriate level of readiness for the information system's implementation, the requirements are then passed to the IS/IT department in the technical level, otherwise, activities that address the shortcomings in the organisational readiness gap should be implemented first, and then requirements passed to the IS/IT department for development and implementation. A new approach for measuring organisational readiness will be introduced in Chapter 10 of this book.

It is noteworthy that the readiness study might result in the implementation of business improvements to fulfil the readiness gap. Such improvements might increase the level of the intended objectives of the project, as a result of the learning process, which in turn affect the requirements specification of the IS/IT. This loop is not shown in the framework for simplification reasons.

The development and implementation phase involves a set of activities;

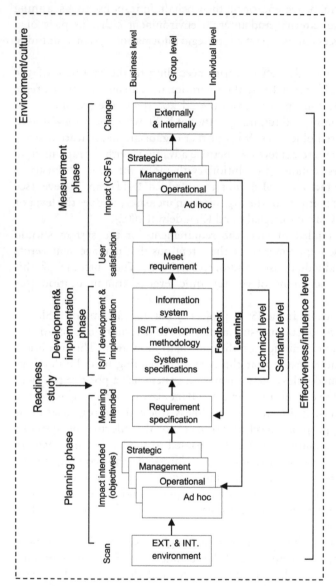

*Figure 3.1* A framework to link success levels to measurement approaches.

producing technical specifications, applying an IS/IT development methodology to deliver an information system that consists of processes, IT components (hardware, software and networks) and people. This could be a bespoke development or an outsourced operation performed under the supervision and coordination of the organisational IS/IT function. At the completion of this stage, the measurement phase starts and success measurement could be performed for each of the three levels of communication.

### 3.7.2 Measurement phase: technical and semantic levels

The output measures for the technical level are concerned with the technical quality of the information system as an output of this level. Measurement of the success at the semantic level is from the user's perspective. Perception is not a precise tool of measurement because it can be influenced by human and environmental factors (e.g., individual characteristics, expectations, etc.) which could affect users' views. User satisfaction is a measure of success in transforming the intended input into comprehensible and satisfactory requirements. User participation and feedback into the development and maintenance activities contribute mainly to the requirements specification step. The feedback arrow in Figure 3.1 represents these user activities.

User satisfaction is considered to be the most used success measure in the IS research field, where it is advocated as a surrogate measure for IS success and effectiveness (DeLone and McLean 1992; Grover et al. 1996; Garrity and Sanders 1998). The reason for this is that many researchers accept the psychological expectancy theory which states that attitudes (i.e. satisfaction) and behaviour (i.e. productivity) are linked. This means that satisfied users will be more productive (Gatian 1994).

### 3.7.3 Measurement phase: influence level

The success measure of the influence level output is the degree of impact or change in behaviour achieved that corresponds with the level of intended impact. The evaluation framework points out that even if a specified level in an organisation is targeted for improvement (e.g. the operational level), the impact could still emerge on the other higher levels.

The degree of impact differs according to the different levels of influence. The greater the difference between the level of intended effect and the outcome level of measurement, the less possible it is to isolate the effect of the information system from other interfering internal and external factors, and the less measurable is the magnitude of the effect the IS/IT accomplishes. DeLone and McLean (1992) state, MIS academic researchers tended to avoid performance measures (except in laboratory studies) because of the difficulty of isolating the effect of the IS/IT effort from other effects which influence organisational performance. This has been echoed in similar terms by many researchers in the IS/IT field (Grover et al.,1996; Garrity and Sanders 1998; Woodroof and Kasper 1998).

This evaluation framework attempts to address this issue by specifying that the measurement of the actual effect of the information system should be the critical success factors (CSFs) for the level of intended effect. If the CSFs are met, then the information system has accomplished its objectives even if those effects do not show on other levels because of the interference of other factor. Information systems objectives are goal-centred which makes CSFs a suitable vehicle for measurement of objectives. Since the objective for the deployment of IS/IT could be of different levels (ad hoc, operational, management, strategic) or on different levels of the organisation (project, group, unit, whole organisation), CSF measurement could be adjusted to attain to those objectives at the different levels of deployment (see Figure 3.2) (Gottschalk and Khandelwal 2002).

In Figure 3.1 learning is accomplished when the experience from previous projects is accumulated and fed back into the planning phase. This is represented by the "learning" arrow.

IS measurement could be performed on different levels in the organisation – individual, group, or organisation-wide (business). This will have an effect on the technical level as it would on the semantic and the influence/impact levels. On the technical level, the type of hardware, software and networks would be different, for example, at the individual level – stand-alone terminals or PC with a stand-alone spreadsheet package; on the group level – having LAN with NT-based applications; at organisation-wide level – having WAN/Internet with servers and applications. At the semantic level, satisfaction would be taken from an individual user, a group, or all of the users in the organisation.

In this evaluation framework, the phases of planning, development and implementation, and measurement are conducted in an atmosphere of common understanding and culture. In order to improve communication (information system components), there needs to be a realm of understanding between management and users. This process is affected by the organisational culture that could be enforced by user participation in the development of the IS/IT. The culture also varies as the maturity of the organisation improves. It will evolve from user participation to having user involvement (Earl 1989). This would be enforced by the IS/IT department by conducting user awareness activities and training programmes (Lynch and Gregor 2001). The realm of culture and experience is represented by the environment box that surrounds the entire framework.

As presented above, measurements are mainly post-implementation events that help in the learning process for subsequent IS/IT projects. Nevertheless, as previously indicated, there is a need for organisational readiness measurement prior to IS/IT development and implementation. This is mainly because many IS/IT projects are either not an outcome of any type of strategic planning or

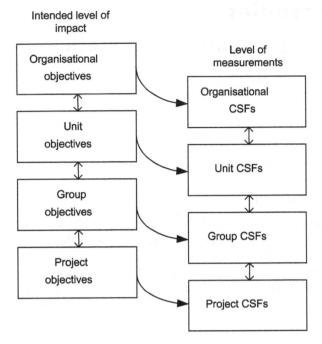

*Figure 3.2* Levels of objectives and CSFs.

they are an outcome of unreliable planning. This readiness measurement is to help ensure that the development and implementation of IS/IT are not undertaken unless the IS/IT has the recipient environment within the organisation which can foster a successful implementation.

# Chapter 4

# Case studies

## Technology vs business approach

### 4.1 Introduction

This chapter presents two detailed case studies: a public sector case and an oil company case. The aim of these two cases is to highlight the importance and the contribution of the organisational factors on the success of IS/IT projects. In each of these two cases a particular IS/IT project was selected and followed through from inception to completion. The analysis will focus on the main organisational factors which were discussed in Chapter 2 and their sub-elements as follows:

1  People (staff, skill, Head of IS/IT department)
2  Process (practices)
3  Work environment (leadership, culture, structure)
4  IT infrastructure (systems)

The first case is a technology-driven one where the project was initiated mainly because the organisation's existing system had reached its limits in terms of data storage and it was no longer able to respond to the organisation's expansion plan. The selection of the international vendor to provide the "total solution" for the organisation was based on its technical capability to design, develop and implement an integrated database for the organisation's applications. The vendor's proposal and thereafter the signed contract did not address the expected changes in the organisation as a result of the implementation of the new system and how to address these changes during the project's phases. This led to a significant disturbance to the work environment and therefore the termination of the project.

The second case is a business-driven case at the early stages of the project but then technology push superseded the business lead. This would almost have led to the termination of the project if top management had not intervened in the implementation process and provided the necessary support for the required changes. The project was first initiated as part of the organisation's plans to improve its performance through better use of IS/IT. Due to the lack of know how (managerially and technically) the project took more than 10 years to be fully implemented with a great deal of disturbance to the organisation's work, associated with exceptionally high costs.

Each case starts with a brief description on the organisation's background and the type of IS/IT projects under consideration. This is then followed by a brief list of relevant sequence of events that took place since the start of the project. The list covers the main activities that the organisation went through to fully develop and implement the system. Within this context, each of the four organisational elements will be discussed highlighting the impact of change on the organisation and on the success of the system.

The two case studies will be further analysed in Chapter 10 in the light of the organisational e-readiness model.

## 4.2 Case study 4.1: ServInst

ServInst is a public-sector organisation that provides services on behalf its stakeholders to wide sectors of investors. The assets owned and managed by the organisation are equal to about US $10 billion and ServInst's yearly revenue is approximately US $1 billion.

The organisation had an excellent history of using IS/IT in support of its work processes. This was due to a strategic decision made by the founding chairman in the late 1970s. Not long after the formation of the organisation in the early 1980s, the existing systems – mainframe based – were built by an international software vendor. Newly hired staff were then trained by the vendor to build the capacity for running and managing the system. The training was system-specific to enable staff to take over the day-to-day operation and maintenance of the system. At a later stage, a large "stand alone" personnel information system was developed, based on an inverted-relational database management system, to accommodate the organisation's 800 employees. That system was used by about 90 per cent of the organisation's staff (more than 700 employees). In the late 1980s there was a complete change of management, where a different team replaced the founding chairman and his team.

### 4.2.1 Sequence of events

- The organisation took a decision to open a new business branch to provide services to the public and relocate the training department and image file-storage to remote sites. At that time the organisation was facing the problem that the computer master-file had reached the maximum record size allowed by the system. This left the organisation with no option but to go for a database environment. For this reason top management decided to replace the main systems with a relational database environment and to build a new network to support the proposed business branch and the new remote sites. The top management decision was influenced by external

sources that many local organisations had successfully converted to database environment.

- The objectives of the new project were to install a network in support
  of the branching plan and the remote sites. Also, to conduct a total resystemisation of all the organisational systems to convert to a relational
  database management systems (R-DBMS) based environment using a fourth
  generation language as a host language for coding the application programs. This resystemisation was also aimed at including the organisation's main legacy systems, the "stand-alone" personnel system based on
  an inverted R-DBMS package.

- The project also aimed at converting all the application programs that
  made up the organisational systems.

- A steering committee was formed of the Chairman's Deputy, IT/IS Unit
  Head, Head of Systems Department, and Head of Operations Department.
  Later, when a vendor was selected and the project started, the vendor's
  team head joined this committee.

- On the request of top management, the IS/IT Unit decided to perform a
  feasibility study where they examined the products of three international
  vendors. They limited themselves to those vendors that could provide the
  organisation with a "total solution", which meant changing all systems in
  the organisation to a relational database environment. It also meant that
  the vendor should have the capability to reprogram/convert all legacy
  applications to the new system environment.

- There were three international vendors:

  ○ Vendor 1 (V1) had a small operation in the country.
  ○ Vendor 2 (V2) had a local branch but did not have the capabilities
    for converting the applications. In order to do so, it would subcon-
    tract the project to many other companies and act only as a project
    coordinator.
  ○ Vendor 3 (V3) had a small office in the country, while it had its regional
    Head Office in another regional country.

- As part of the feasibility study, the IS/IT Unit produced criteria for vendor
  selection (mainly technical):

  ○ the DBMS product had to be a full relational DBMS;
  ○ the development should include CASE tool for ease of design;
  ○ the DBMS had to support the use of credible fourth generation
    language;
  ○ training had to be part of the whole development contract allow-
    ing for technology and experience transfer from the vendor to the
    organisation.

- After the feasibility study was concluded, the organisation set up a team
  consisting of the IS/IT Unit Head and a full-time consultant who was hired

to serve in this team. The team was formed with the following terms of reference:

- To produce a general system specification based on the "total solution" approach and a report for proposal (RFP). This needed to be done while having in mind the at-most utilisation of the existing organisational IT infrastructure, especially the hardware.
- To recommend a vendor to conduct the detailed requirement specifications which was termed "strategic study".
- To oversee the detailed requirement specifications of the "strategic study".
- To recommend a vendor out of the three vendors specified by the feasibility study to develop and implement the system.
- To form the team to represent the organisation in the project's development team.

- After several meetings with the vendors' representatives, the initial recommendation of the team was made in favour of V3. This decision was based on the following.

  - V1 was much more expensive than the other two vendors. It also had no support office within the country, and its product was an "inverted-relational" not "full-relational" DBMS.
  - V2 was expensive, and it would subcontract the project to many vendors which could cause problems in the development, coordination and support. Also, its CASE tools were largely customised to support the methodology used by only one of the subcontracting vendors.
  - V3 had the most reasonable price, and its product had started to gain a good reputation worldwide. Also, the company was planning to establish an office within the country that could provide after-installation support.

- It was agreed that the final selection decision should be made after the completion of the "strategic study" to ensure that the vendor had the capability to deliver the specified systems. The team had requested V3 to conduct the "strategic study". This request was made independently from the main project's contract for the delivery of the final "total solution". It is interesting to state here that the initial study which was carried out by the IS/IT Unit favoured V3; if the strategic study was to be performed by V3, which was the favoured vendor, then some of the output of this study (e.g. data entity relational diagram) could be directly incorporated in the development method which would save the organisation time and money, which could have incurred had they selected any of the other vendors. In fact, an informal decision was made to select V3, but for formality reasons, which include government regulations, this process had to be followed.

- The "strategic study" was conducted by four V3 employees and spanned over four months.
- The study included interviews with all users' representatives, especially managers and key persons. Questionnaires were also used in addition to reviewing the existing documents.
- The outcome of the strategic study provided the organisation with the following information:

  o  the resources required for the project: staff, time, hardware, etc.
  o  documentation of all work processes: such documentation did not exist
  o  determining the interaction between different systems
  o  information regarding the condition of data-files, formats, storage media types, size, backups, etc.
  o  finding opportunities to improve business processes and structure
  o  matching the project's requirements to the capabilities of potential products and vendors
  o  a data entity relation diagram

- At the end of the strategic study, the organisation officially announced that V3 was to provide the R-DBMS and to develop the applications needed in fourth generation language. A contract was signed to implement the "total solution" in two years.
- Upon the request of the organisation, the vendor provided the organisation with their team members' names and résumés and these people were interviewed and approved. This approval process was not stated in the signed contract, but was agreed upon by the two parties verbally.
- The two project teams from both the vendor and the organisation worked together in a partnership to provide the joint project team but each had their own leader. The IS/IT Head led the team from the organisation side.
- Because the requirements were defined in the "strategic study", the joint project team immediately started the work on the design of the system.
- For the first six months, activities went according to plan. After that and while still in the design phase, the vendor started to change its team members, reallocating the experienced ones to other projects and replacing them with members with less or little experience. The reallocation and replacement of the vendor staff continued throughout the project.
- On the organisation side things did not look much better. Both IS/IT Unit and users, demanded that the key staff assigned to the implementation of the project in the early stages return to their original posts. They claimed that the day-to-day work was negatively affected by their full-time assignment to the project. Top management caved in, and most of those who left the team were replaced by junior, less experienced members.
- In spite of the low level of experience and skill of those junior staff, no training was provided and they were left to learn the basic skills on the job.

- As the project progressed and the new system output started to appear, user resentment and criticism started to grow. No effort was made by the project team to clear possible misunderstandings or increase awareness regarding the project's aims, objectives and implications.
- As the project progressed in time, user rejections and reservations gradually gained top management sympathy, which in turn started to influence their support for the project. This negatively effected the required approval for changes and modifications in system design, organisational structure and work processes that would come up during the development phase. Such changes had not been planned for at the start of the project.
- By the end of the project duration users rejected a large number of the system's functionalities. Users did not feel that the functionalities had fulfilled the needs of the work tasks.
- Both sides, the vendor and the organisation, agreed on an extension of the contract with amended penalty terms set by the organisation. This was followed by a second extension with no acceptable results.
- Four years from the time it started, and at a cost of between US $7 to 10 million, the project was declared a failure and the matter was handed to the courts.

### 4.2.2 Analysis and discussion

This section analyses and discusses the above unfolding events in terms of the organisation's main factors: people, process, work environment and IT infrastructure. The people factor is discussed in three categories: staff, skills and Head of IS/IT. The process is addressed by the organisation's practices while the work environment factor is addressed by leadership, culture and structure. Finally the IT Infrastructure is addressed by the available systems.

### I People: staff

- After withdrawal of the experienced staff, the vendor provided inexperienced staff to the project.
- There was not enough qualified staff in the organisation who could handle both the ongoing daily organisational work and the new replacement of the systems. Both of those tasks needed total commitment.
- The organisation decided to remove its key people from the project team and return them to their original departments to do their daily tasks, while assigning staff with little or no experience to the project.
- The junior staff assigned to the project team had no training prior to, nor during the project.
- Because of the complexity of the project and the need for a holistic view of the integrations between the many different sub-systems, training of the inexperienced staff assigned to the project was highly unattainable. This is

because the current systems were not sufficiently documented, leaving the knowledge and the know-how residing only in heads of key people. This was complicated by the nature of the culture existing in the organisation, namely an individual un-cooperative culture. Staff tried to keep their knowledge and experience to themselves for different reasons including job security. Key people considered that keeping the knowledge to themselves would make them indispensable and increase their value to the organisation.

- Two years from the start of the project, the project leader on the vendor side was changed and the new one had a different approach to the project's development: he attempted to change the new system design.

## 2  People:skill

- The contract signed with V3 included up to US $70,000 worth of training.
- The contract also included technology/experience transfer by having organisational staff work alongside the vendor's staff.
- The plan was that as the project progressed through the development life cycle phases the proportion of organisation staff would increase and the vendor's staff decrease (vendor's/organisation: 80–20 per cent in phase I, 60–40 per cent in phase II, 40–60 per cent in phase III, 20–80 per cent in phase IV). This would reach the situation where each of the vendor's staff members would work with and train four of the organisation's staff.
- The project leader on the organisation side (IT/IS Head) noted that he did not have the needed skills for such a major project. He did not receive proper training and had no previous experience.
- The organisation assigned inexperienced staff to the project, while the project needed the best skill and experience. This was true with both the IS/IT Unit and user groups.
- The vendor's experienced staff were withdrawn after six months. Many of the vendor's new staff who were assigned to the project had little experience.
- In many instances, especially in the advanced stages of the project, the organisation's staff were more experienced than the vendor's staff who were supposed to train them. On a few occasions, the organisation requested the removal of some of the vendor's staff because it thought they were incompetent.
- The skills of the organisation's IS/IT staff on the project team were mainly programming with limited systems analysis experience. Very limited number had the skills to use the inverted relational DBMS and its query language.
- Training was largely unavailable for the inexperienced staff who worked on the project. The organisation, at some time during the project, provided the staff with video-based lectures to be viewed in the employees' own time. This did not result in a considerable success.

- The vendor did not suggest any training or pre-project skill requirements for the project team members on the organisation side before the start of the project.

## 3 People: Head of IS/IT

- IS/IT Unit Head and the project leader had a middle-management status.
- The Head of the project did not have enough authority to implement changes and modifications needed mainly in the organisational structure and processes.

## 4 Work environment: leadership

- The leadership support to the project was strong in the beginning, but as time went on the support became weaker.
- The prolonging of the project negatively affected top management's support.
- Top management's attitude towards the project fluctuated according to the external and internal pressures. For example, if the media focused on a certain aspect of the project that aspect received priority over others. Also, internally spread "rumours" might affect the amount of support given to a certain part of the project.
- Top management gave priority to the daily operation of the organisation over the project.
- Top management relied on acquiring their IS/IT related knowledge from external sources and largely from what other organisations were doing.
- Even though the organisation was heavily dependent on IS/IT in its day-to-day operation, top management considered IS/IT as a tool necessary for smooth functioning. The heavy dependency on IS/IT should have called for a strategic view of IS/IT and its consideration as an asset to the organisation.
- At the early stages of the project where the outcomes were not visible, top management support was high. As the system development progressed, users started to see the outcome and top management started to receive many complaints and rumours which caused support and enthusiasm to lessen.
- The organisation's incentive approach was individually based not team-based, and did not support knowledge-sharing and cooperation.

## 5 Work environment: culture

- There was a lack of positive relationships between different groups/departments in the organisation. Despite the fact that work processes cut through departments' boundaries, there was no cooperation between them.

Also, within each group, the individualistic attitude kept employees from sharing knowledge.

- Different managers were suspicious of the project, thinking it was a plan being executed to deprive them of their powers in favour of others. This caused them to resent the project and not to cooperate with the project team.
- Key staff considered keeping knowledge and experience to themselves as a job security tool, and were not forthcoming in cooperating with the project team.
- The lack of decisive leadership by top management allowed conflicts between different departments in the organisation to increase. Also, security and confidentiality issues were obstacles in adopting the incremental approach in building the systems that was preferred by the vendor. The project team was prevented from introducing incremental changes. Top management did not allow the team to "touch" the current work practices when the team saw fit, even though it was covered by the contract to convert it to the database environment.
- There was a lack of strong support from top management for the necessary changes in the processes and structure.
- The vendor did not suggest any changes in the culture, and in fact, the issue of culture was never mentioned.
- Some of the key users felt threatened by the new system because it would automate some of the tasks they believed would give them advantage if they remained un-computerised. An example of this is a key employee who memorised all of the organisation's account numbers to be filled in on forms. He resisted the automation of the process.
- The project team did not make an effort to relieve the tension in the relationship that occurred within the organisation because of the introduction of the project.
- The project team, at some point in the project, adopted the user-centric approach in testing the system's functions. The key users at that time became a burden on the project by being negative and demanding all the time.

## 6 Work environment: structure

- Even though in the organisational chart the IS/IT Unit has a senior status through the "technical office" of the "Automation Sector", the new top management, unlike the previous one, left the position empty for over a decade.
- The Head of the IS/IT Unit had a middle-management position.
- Even though structure changes were obviously needed, as implied by the IS/IT Unit Head, the vendor did not suggest any changes in the "strategic study".

## 7 IT infrastructure: systems

- There were problems with the format of the data-files in the organisation. The size of the master file reached the maximum allowed by system.
- Data that were used came from external entities and were written in a non-compatible format that required special and time-consuming conversions. Also, such data were not reliable, with bad data both in context and in content requiring editing procedures.
- The new system's design did not attempt to resolve the ownership of the data within the organisation.
- The vendor treated the project as if it were building a new system in an organisation with no prior systems. The new design changed all the interfaces and environments that users were using, causing considerable resentment amongst users.
- After two years in the project, the vendor changed the system design, which caused unease within the project team.

## 8 Process: practices

- Processes were neither defined nor documented prior to the project.
- No process change was suggested by the vendor.
- No process changes were allowed by top management and users when the needs were recognised by the project team.
- Because of the inability to re-engineer processes, "corner-cutting" and "go around" techniques were used, such as allowing two units to access, modify, and delete the same data. This caused problems and conflicts for the project team with and amongst user groups.

### 4.2.3 Summary and findings

- The selected vendor used the project to train its own new and inexperienced staff.
- Organisations need to clearly state in the contract that the qualifications of the vendor staff, including latecomers, should be approved prior to commencing on-site. In this case the vendor had a high turnover of staff. Experienced staff approved by the organisation were assigned to the project for a short period of time, at the beginning of the project. Many of the vendor's new staff who joined the project at a later stage had little experience, and many of those said on occasion that they were "filling time between jobs" until they found a better job.
- It is important to focus on the assigned staff for the project when assessing the organisational factors. There might be competent staff in the organisation but they might not be allocated to the project.
- The vendor asked for a key person for each of the users' functions to

participate in the system testing. This made the approval subject to this one person's prejudices and reservations. It might have been different if the approval had been performed through a team representing each user's function.

- Lack of positive relationships between different departments in the organisation caused resistance to the required change in structure and processes.
- Because key staff considered keeping knowledge and experience to themselves as a job security tool, they were not forthcoming in cooperating with the project team. This was complicated by the almost complete absence of systems' documentation, and the little documentation that existed was obsolete or not comprehensive.
- The individualistic culture almost entirely prohibited the existence of inhouse training programmes for new and junior staff by senior staff. Those programmes that were rarely conducted were largely ineffective because they lacked the practical know-how and documentation.
- The lack of decisive leadership allowed the conflicts between different departments in the organisation to increase, having a negative effect on the project success.
- The lack of strong support from top management prevented the project team from implementing necessary changes in processes and structure.
- The project team did not make an effort to relieve the tension between users caused by the introduction of the project. They should have introduced a user awareness programme to explain the benefit of the project and worked towards eliminating unfounded suspicions.
- The project leader thought that the vendor's main concern at that time was to win the contract. Therefore, the vendor stayed away from highly sensitive areas such as culture, processes and structure that might have caused it to lose the bid.
- Because top management caved in under pressure from both IS/IT Unit and user groups, the project team was not given the authority to modify the organisational structure when it was obviously necessary to solve some of the design problems that appeared later in the project.
- The new system design was not successful in resolving the ownership of the data within the organisation. This was an issue that caused user resistance to the project.
- The vendor treated the project as if it were building a new system in an organisation with no prior systems. The new design changed all the interfaces and environments that users were used to. This caused resentment amongst users. The new design was understood to imply that users ignore their experience and knowledge gained over the years. This knowledge and experience was considered by many key users as their "value" in the organisation. It was perceived by many users that they would be forced to start afresh if this system were to be implemented. This problem is more acute than it may seem. Most of the people and especially the key staff on both

the IT/IS and user sides knew of IT/IS only through the current old system, and at this organisation only. They had not experienced IS/IT in any other context. Almost all of them had no IS/IT education and came from non-technical disciplines and backgrounds. They had experienced IS/IT and been trained only in this organisation and on its system. Most IS/IT personnel did not even own computers at home. Working with IS/IT for many years at one of the pioneering organisations in introducing IS/IT in the country became a personal prestige that produced a social status in addition to being a skill and a career, especially for those in the IS/IT Unit who carried the "Systems Engineer" title. To think that the new system would deprive them of all of the status they had gained and return them to square one by becoming trainees instead of experts was a cause of strong resentment felt by almost all the key staff.

- The "strategic study" did not address the real problems with the old system. For example, the user managers had to answer a questionnaire with multiple choice answers regarding enhancing performance of the work task. Many of the performance indicators were not applicable to the actual situation. Those indicators were based on the vendor's own experience.
- The withdrawal of the vendor-approved experienced staff had negatively affected the organisation's trust in the vendor. The organisation had not addressed this issue in the contract and took the vendor's initial agreement as being enough of an assurance. Also, the organisation assumed that the vendor would do its best for the project to succeed because it would give the vendor a good reputation in its local market.

## 4.3  Case study 4.2: OilCo

OilCo is a major oil company which was established in the mid 1930s. As part of major rehabilitation activities in the 1990s, the company set out ambitious plans to improve its performance and operation including the IS/IT project, which is the subject of this case study.

### 4.3.1  Sequence of events

- As part of the upgrading process of the company's infrastructure, an international oil consultant rated the state of OilCo's data as 2/10 of how it should be. The Chairman and his deputy expressed the need for a project that would resolve this serious problem.
- Various feasibility studies for a data management project were solicited from different international vendors. Studies were conducted to produce proposals for tender almost a year later.
- An organisation-wide communication network of a general-purpose nature was implemented to set up the IT infrastructure. A major database

software was chosen to be the "backbone" relational DBMS for OilCo applications.

- Three years later, a data management (DM) steering committee was set up to oversee DM project implementation.
- In the early days of the DM steering committee, members unofficially favoured a particular software vender, which was subsequently selected, for the OilCo's data management system. This was mainly because the members knew that the vendor was used by their competitors.
- Subsequently, this vendor (V) was asked to conduct a new feasibility study for the project.
- The vendor was finally selected later in the year for the implementation of the company's integrated data management system.
- The integrated data management project began immediately after the contract was signed. A joint implementation team was set up first under the finance department of OilCo then it was moved to a technical department.

### 4.3.2 Project objectives

- The main objective of the Data Management Project was to migrate multi-disciplinary data related to production, reservoir, geologic, geophysical, petroleum, drilling and surface facilities from several legacy systems, hard copies and tapes to a modern, secure and robust integrated database system using a commercial package. Easy data access to the integrated database was expected to greatly enhance staff productivity. An internal study had shown that the company's geoscientists spent about 35 per cent of their time on data search (hardcopy, diskettes, Excel sheets, Access sheets, legacy applications), reformatting and re-organising. This project was to cover the data for all OilCo's oil-producing areas and to make the data accessible to professionals.
- The new system was to provide a wider accessibility to management through Internet browsers.
- Selective oil production applications were to be linked with the new system.
- Links were to be established from oil production operations to the new system.
- Enabling the implementation of DM projects for other groups in the company.
- The new system was to be aimed at 500 users on an organisation-wide network.

### 4.3.3 Implementation of the project

- Implementation of the project was planned in three phases over 30 months (see Figure 4.1). Table 4.1 briefly explains the aims and objectives of the development and deployment phases.

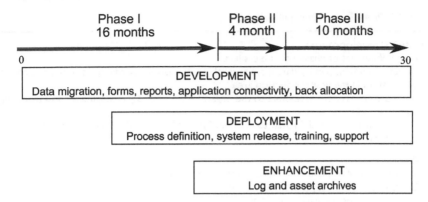

*Figure 4.1* OilCo's project phases.

- A joint OilCo/vendor alliance team was to implement Phase I of the project with the Vendor executing Phase I for one production area and to provide training. OilCo staff, on the other hand, would conduct almost the same steps in Phase II and III for other production areas, where the vendor would only have an advisory role during Phases II and III (Table 4.1).

- At this stage of the project, almost five years after the initial feasibility studies were conducted, the vendor's role was identified as follows:

  ○ Initiation of the project environment which included project management and control procedure, creating project plans, index and list of the digital data to be loaded, conducting a detailed user requirement and setting up procedures for data acquisition.
  ○ Loading of historical data.
  ○ Setting a convention mechanism for transferring data to other application programs.
  ○ Loading of the new data generated in Phase I.
  ○ Training of the organisation's project team.

- After six months on the project a number of problems surfaced. For instance the contract did not state what would happen after the data were entered into the database or how users would be connected to the system and whose responsibility it would be to train them. It did not even specify whose responsibility it was to install the needed hardware for the users to be connected to the system. Also, it did not state whose responsibility it was to fetch, collect and capture the old data from different sites in the organisation and clean and convert them to a readable format for data entry in the new system. Such problems caused delays and budget overruns that almost brought about the termination of the project (Table 4.2). Because of these difficulties a system implementation team was formed entirely from OilCo staff to overcome these problems.

*Table 4.1* Project's aims and deliverables

**Phase I (Development)**

Aim and objectives of Phase I of the project

- Migrate multi-disciplinary data from several legacy systems, hard copies, and tapes to the new system in the first production area
- Make the new system accessible to the organisation's managers, engineers and technical supervisors
- The vendor to train OilCo staff to conduct Phases II & III of data classification and entry for all production areas

**Deliverables**

- Migrate production data from legacy databases of the first production area
- Generation of monthly production reports
- Applications connectivity to the new system

**Phase II (Deployment)**

Aim and objectives of Phase II of the project

Deploy the DM system to end-users by:

- Installing adequate user-oriented hardware, software and network for the new system to go live
- Identifying and changing business and dataflow processes for capturing, validating and storing all pertinent data
- Providing tailored training programme based on users' needs and disciplines
- Supporting the system with a skilled DM team to operate maintain, and manage the "live" system
- Providing and allocating DM coordinators within user groups on site
- Conducting awareness programmes to overcome users' sceptical and reserved attitude towards the system

**Deliverables**

- Effective data workflow and service lines
- Validation and management of data
- Generation of daily task data and reports
- Web-enabled data query and access
- Export/import of data to/from other applications
- System and database backups to be run frequently

- Even though it was not planned for, but for the purpose of overcoming some of the problems stated above, the deployment phase of the project was started six months after the beginning of the project and in parallel with the development phase.
- A deployment team was formed in an ad hoc fashion, bringing people from different departments in the organisation who, in most cases, had no previous knowledge and experience related to the project.
- The company went into top management change at about mid-life of the project.
- At the end of the project, the project team started an awareness programme

*Table 4.2*  Difficulties faced in project

---

Difficulties faced in the Development Phase
- Data preparation took longer than expected because the old data were spread around the company's departments in different forms with no unified filing system, and sometimes no filing system at all.
- Administration-related difficulties like visas for staff and high turnover of the vender's people, where some of the qualified people were transferred to other projects and new inexperienced staff were brought in.
- Some of the user requirements were not clear regarding, for example, forms and reports. Also, the development was based on existing business and dataflow processes.
- User feedback was low due to the unenthusiastic user's attitude towards the project.
- Many of the project team had to be trained on basic IS/IT skills during the project's development. They were expected to have such skills as those of Unix and managing large databases.
- Also, many of the team members did not have full-time commitment to the project in the initial stages.

Difficulties encountered in the Deployment Phase
- The deployment team faced difficulties with users' acceptance of change in their work process.
- The team also faced uncertainty on the time and resources required for the completion of its task.
- The project spanned over several departments and because of the lack of flexibility in changing organisational structure, the structure remained as it was even when change was needed.
- Lack of obtaining support of top management to progress smoothly. The team also had difficulties with regard to user co-operation throughout the deployment activities.
- Also, the lack of experience among most of the team members caused difficulties and delays in performing the scheduled tasks.
- Many of the deployment team had to be trained during the project with the basic IS/IT skills.
- Also, many of the team members did not have full-time commitment to the project because they were assigned to other tasks in their original departments in the organisation. Most of the members were assigned to the project team on a part-time basis until advanced stages of the project, when top management assigned them fully to the project.

---

targeted at users and focused at user managers to promote the use of the system.
- The project started to run with only a few licences from the Vendor to be used by the project team members, while the hardware was a small server with little capability.

### 4.3.4 General remarks

- OilCo did not have the knowledge and expertise to make a decision regarding the choice of the database system needed, which contributed to an eight year delay from starting the feasibility studies. According to an OilCo senior manager "The decisions taken regarding the data management project were not made properly".

- The project was not developed as the vendor proposed in the feasibility study. No consideration was given to the preparation steps in the actual contract with the vendor. In the initial study, the vendor recommended some pre-steps to streamline business processes and OilCo's old top management did not approve them for financial reasons.

- The project team recognised part of the problems at an early stage in the development phase after the start of the project. Some measures were taken to overcome them, but because of lack of resources, i.e. knowledge and real top management support, the problems could not be resolved in time. These problems had negative effects on the success of the project.

- Both the project leader and the implementation team head admitted the problem that the system was not being used by users in the way it ought to have been.

- The usage of the system after 4 years of implementation was around 40 per cent of the expected usage. Users still relied on their old ways to conduct 60 per cent of the work tasks.

- The problem caused by the decision of OilCo's old top management not to approve the recommended pre-steps to streamline the business process produced by the consultant was compounded by the vendor's acceptance to submit an alternative proposal focusing on technical implementation issues only.

- The final proposal by the vendor did not address what to do next after loading the historical data. It did not address the needed hardware and software licences, training, processes, organisational structure, skills, user and management issues, etc. The proposal even failed to address some issues prior to loading the historical data, such as fetching, collecting and cleaning the old data.

### 4.3.5 Analysis and discussion

This section analyses and discusses the above unfolding events in terms of the organisation's main factors: people, process, work environment and IT infrastructure. In a similar manner to the previous case, the people factor is presented in three categories: staff, skills and Head of IS/IT. The process is addressed by the organisation's practices while the work environment factor is addressed by leadership, culture and structure. Finally the IT Infrastructure is addressed by the available systems.

## I People: staff

- The internal project leader was appointed only a month prior to the official start date set by the contract, when he was allocated a team of thirteen partly dedicated personnel. Almost all, including the project leader, had no prior experience in implementing IS/IT projects.
- During the project, major changes in staff occurred and the team ended with twenty-five members who were mostly different from those appointed at the start of the project. This was due to the hiring of new staff as a result of the support given by the new top management to bringing in more qualified staff.
- The project team was not fully dedicated to the project in its initial stages. This was true for both the development and deployment teams. This problem was mostly resolved towards the end of the project.
- The project suffered from lack of adequate human resources, in terms of quantity and quality, until the late stages of the project life.
- The vendor staff kept on changing throughout the project life.

## 2 People: skill

- The project started with levels of staff skills lower than it actually required to be successful. The project team did not have the minimum needed skills and received no relevant training prior to the start of the project.
- The vendor recommended that prior to commencing the project, the team should acquire at least some skill in dealing with UNIX and management of large databases. This recommendation was not adhered to.
- Training of the project's development and deployment teams was conducted during the project.
- There was also a lack of adequate skills on the vendor side. About 50 per cent of the vendor staff were inexperienced.

## 3 People: Head of IS/IT

- At the beginning of the project, the head of the project team was appointed from the administration division as it was considered that the new system was supplying administration services.
- The new management thought that this project was of a technical nature, so the team was moved under a technical-oriented division in the company and was upgraded to become a group and the project head became the Data Management Group Manager.

## 4  Work environment: leadership

- The old management did not want to pay for the full proposed solution (which included process change) in the beginning and informed the project team that they could ask for more money at a later stage of the project, even though it was known that the budget-approval procedure might take more than a year.
- Management's view on the project was changed by changing the top management personnel which occurred about half way into the project's life. The situation was improved as described by the project Head: "This situation has significantly changed in that whenever we raised a budget for the project they approved it without constraints."
- Top management were kept informed on the project's progress by a monthly newsletter along with monthly, quarterly and annual progress reports and by regular feedback from the project team manager during senior management meetings.

## 5  Work environment: culture

- Users were not involved in the project at the beginning. They were not involved in requirement specifications and process change decisions.
- The organisation suffered from lack of trust and communication between the project team and user groups. This was manifested in the users' spreading of what the project team felt as "false rumours" regarding the capabilities of the new system to top management.
- The project team did not consider the culture issue until it became obvious that it was an obstacle to the success of the project and, even then, they did not know how to deal with it until the project was almost completed.
- At the end of the project, the team started an awareness programme targeted at users, and focused on user managers to promote the use of the new system which was mostly ignored.
- The team also allocated support staff at user sites and had a help desk along with Internet project page for fault reports and inquiries.
- After the execution of the above steps, the use of the system went from 10 per cent to 40 per cent in three months.

## 6  Work environment: structure

- The project team was first allocated within a division that was supplying administration services. The head of that division did not fully understand the team's needs and requirements. The new top management recognised this issue and moved the team under a technical division.
- At the beginning of the project, the management did not position the project team within its organisational structure therefore it lacked the

required authority to be taken seriously enough when interacting with user groups and management.

- The internal structure of the project team was formed at the beginning of the project according to the vendor's recommendations, and it continued until the deployment stage started where the structure of the team changed. This was done because the needs at the deployment phase were different to those at development phase.
- There was no flexibility in the organisational structure. Changes were very difficult to implement. As a result, no restructuring was performed even on the user side, when it was seen to be necessary.
- Close to the end of the development phase, the new management were convinced that the project team should be a group by itself. The project team was given more authority, where the project leader was promoted and it became part of a technical division. This gave the team more power and started to demand more co-operation from user groups.

## 7  IT infrastructure: systems

- The company had an organisation-wide communication network based on the UNIX network platform. E-mail and Internet were used extensively within the company as part of day-to-day job tools.
- The new system's database package uses R-DBMS environment in implementing OilCo's data management system.
- The project team members started the project with one small server and little hardware capability. At deployment, the team recognised that more licences and a better server with bigger storage and power were required.
- The deployment team had to struggle with the shortage of hardware which caused some delays until extra budget was approved, because the old management did not approve enough budget at the beginning. This was changed later with the new management.
- The project database was configured around two servers. One server was used for the live deployed system and the other for development, maintenance and backup purposes. System and database backups were run daily, weekly and monthly, and the backed-up tapes were kept off-site to provide data security and to ensure disaster recovery capability.
- Users could connect to the servers from their PCs and access the new system. The system offers users the following utilities in an on-line environment:
  - shorter cycle of dataflow
  - on-line accessibility with improved data quality
  - automated report generation
  - monthly production reports
  - a data management system loaded with all the historical and up-to-date production data

- connectivity with other applications and other corporate databases
- management could browse the system's database via the Internet

- New attempts were made, during the deployment stage, to develop a decision support system based on the new system.

## 8 Process: practices

- The vendor recommended conducting business process re-engineering (BPR) at the feasibility stage, but the management favoured focusing on the system deployment only.
- During the deployment phase, it became necessary for the organisation to start changing its processes. The project team had to change the processes in the way of capturing data and retrieving /disseminating them to match the new system's requirements. At this stage, the policy of the organisation was "whoever captures/generates the data is the one who should enter them in the system". The more understanding of the processes the team acquired, the more they saw the need to modify/change the processes.
- Although the vendor tried to perform process changes, the project team did it themselves, where they admit that they "struggled a lot in doing it". However, they claimed that they "did it successfully and gained experience by this". They also thought that they were capable of conducting BPR better than an external consultant because they "understand the culture and political issues in the organisation better than the consultant", which they believed would affect the success of BPR.
- BPR or process changes were conducted to fit the new system. Those changes took place after the system development phase was completed.
- In order to fulfil the new requirements, the process change targeted the way data were captured and distributed. This also covered the responsibility for producing, entering or manipulating data between and among various departments. This affected all classes of data and whoever was responsible for them.
- The changes in processes affected both the main/generic processes as well as the smaller specific sub-processes. Responsibilities of some processes were shifted to more relevant departments.

### 4.3.6 Summary and findings

- The project did not address the preparation of any of the organisational factors before the project began. Required skill, culture, leadership, staff, structure, systems and Head of IS/IT were not prepared for the project, nor the need for BPR or process changes.
- During the project the project team attempted to fill the gaps and fulfil what should have been performed as pre-steps before the start of the

project. These attempts continued even after the project had ended. The team achieved some progress but still not enough to make the project a success. The project team turned out to be more like a rescue mission, trying to implement some of what should have been implemented before the beginning of development of the system, such as improving relationship with users.

- IS/IT awareness amongst the old management of the organisation was low. It seemed that the success of the project for the management was not as important as the prestige of having this "vendor's product" at the organisation. This was mainly because the system was already being used by their major competitors in the region.

- The vendor's final proposal did not address what to do next after installing the system and loading the historical data. It did not address the required hardware and software licences, training, processes, organisational structure, skills, user and management issues, etc. The proposal even failed to address some issues prior to loading the historical data such as fetching, collecting and cleaning the old data.

- The new management became interested in having this project succeed and they gave it considerable support. This resolved the leadership problem.

- Also, the awareness programme targeted at top management helped in this regard. The case shows that the timely change of top management contributed to the survival of the project. Had the attitude of the old management remained as it was at the beginning of the project, the project would have been terminated.

- The improvement on the culture issue had an effect on the usage of the system. The execution of the awareness and support programme improved communication between the project team on one side and the users and management on the other, making usage of the system increase from 10 per cent to 40 per cent in three months.

- The organisational structure seems to be tied with the culture. Decentralisation of control would not resolve the problems regarding the system use, especially when there still exist communication problems between project team and user groups.

- Since the project team was not experienced in process improvement or BPR, the improvement in the process domain was slow. The team attempted to improve the process by following a "trial and error" approach.

# Part 2

## Enablers

### Learning organisations and strategies alignment

# Chapter 5

# Developing sustainable IS/IT capabilities

## The learning organisation

## 5.1 Introduction

IS/IT investment is being increasingly used in support of business objectives. The success of these projects cannot only be measured by the successful installation of the system, i.e. hardware and software, nor by the level of users' satisfaction. A fully operational system with satisfactory functionalities can be alienated and not be used by employees. The successful implementation of IS/IT in an organisation is highly related to its ability to absorb and integrate the proposed systems into its current practices. Systems can only achieve their intended business objectives if they are fully integrated into the organisation's current work practices and are accepted and supported by both employees and management.

The two critical elements that can significantly influence the level of IS/IT project integration into organisations' work environments are:

1 *process alignment*: the ability to align the organisation's processes with the proposed system's functionalities, whether it is a commercial system with embedded "best practices" or a bespoke system with newly adapted processes. (In Chapter 2, the two-way influence of process and IT on busiess improvements was discussed.)
2 *people*: the ability of employees to accept and adapt to the new system quickly and efficiently.

These two elements are determined by the level of a) awareness among management and employees of the benefits (operational and strategic) that IS/IT projects can bring to their organisations, and b) their experience and know-how of how IS/IT can be used to realise such benefits. These two issues are underpinned by two organisational capabilities:

1 The presence of a work environment that supports, motivates and empowers employees to innovate and seek improvement changes in line with the business objectives.

2   The availability of advanced IT infrastructures that allow employees to develop appropriate levels of experience and understanding of the potential of the new system to improve the performance of their current business processes and to achieve their department's or corporation's goals.

In order for an organisation to achieve the required level of capability, it has to carefully develop and implement plans to create a work environment where innovation and continuous improvement can thrive. The focus should be on developing and sustaining a highly skilled and flexible workforce which will have the skills and competencies to continuously introduce improvements through better and more streamlined business processes which are enabled by advanced IT. In this context, organisational learning and knowledge management become a necessity for organisations to sustain business improvements and competitive advantage out of their IS/IT investments.

## 5.2 Organisational learning

Theories of organisational learning owe much to the work of Argyris, Schön, and Senge. Argyris and Schön introduced the concepts of "single-" and "double-" loop learning, while Senge developed the ideas of "adaptive" and "generative" learning (Argyris 1977, 1992; Argyris and Schön 1978). Single-loop learning simply involves the correction of errors through a feedback loop. This is very similar to Senge's concept of adaptive learning which centres on evolutionary changes in response to developments in the business environment which are necessary for a survival of the organisation. Such learning does not deliver competitive advantage but it is essential for an organisation to survive. Double-loop learning is cognitive and goes beyond the provision of immediate solutions to current problems. It involves the development of principles that may inform and determine future organisational behaviour and lead to new ways of doing business (Argyris and Schön 1978; Argyris 1992). Generative learning is concerned with building new competencies, or identifying and creating opportunities based on leveraging existing competencies.

Organisational learning is the process of continued innovation through the creation of new knowledge (Quinn et al. 1996; Nonaka 1991). It is an ongoing process that takes place as employees engage in knowledge work (Davenport 1998). It involves two types of knowledge.

1   *Explicit knowledge* is knowledge that can be codified. It can be expressed in formal and systematic language and shared in the forms of data, scientific formulas, specifications and manuals. It can be processed, transmitted and stored relatively easily. Thus, it is easier for organisations to capture and make available to all members of the organisation.
2   *Tacit knowledge* is highly personal and hard to formalise. It is deeply rooted in action, procedures, routines, commitments, ideals, values, and

emotions (Cohen and Bacdayan 1994). It resides within the individual who has a wealth of tacit knowledge deeply rooted in his/her actions, and his/her commitment to a particular craft or profession. In most organisations, tacit knowledge is rarely shared or communicated. Therefore, it is often lost when the individual possessing it leaves the organisation.

Nonaka (1991) states that organisational learning emanates from the iterative process of articulation and internalisation. Articulation occurs when an employee's tacit knowledge is captured as explicit knowledge and internalisation occurs when this captured explicit knowledge is transformed into another employee's tacit knowledge. Therefore, organisational learning occurs at the intersection of tacit and explicit knowledge during the interaction of the various employees, departments or teams in a firm (see Figure 5.1).

## 5.3 Competitiveness and organisational learning

The ever-increasing interest in knowledge in recent years has been accompanied by a renewed discussion of organisational learning and knowledge management and, more specifically, the potential for an organisation to generate competitive advantage on the basis of its knowledge assets. Harvey and Denton (1999) put forward several reasons for this including:

- the shift in the relative importance of factors of production away from capital towards labour, particularly intellectual labour;
- the ever more rapid pace of change in the business environment;
- widespread acceptance of knowledge as a prime source of competitive advantage;
- the greater demands being placed on businesses by customers;
- increasing dissatisfaction among managers and employees with the traditional, command and control management paradigm;
- the intensely competitive nature of global business.

The volatility of the environments in which organisations operate has made the creation and sustainability of competitive advantage an even more

*Figure 5.1* Learning environments.

demanding task. Equally, the recognition of knowledge as the single most important source of competitive advantage has developed alternative ways in which organisations can build and sustain superior performance. It is now possible for organisations to achieve greater flexibility and adaptability through continuous organisational learning and the improved management of their knowledge assets on which their core competencies are based (Pemberton and Stonehouse 2000). Thus, in an increasingly competitive environment, focusing on organisational learning and knowledge management is seen as a critical route for rapid development and effective use of knowledge assets that are superior to those of competitors. In summary, organisations that learn quicker than their competitors, and as a consequence deploy their knowledge assets more effectively, are better placed to create and sustain a competitive edge.

One of the most important roles of organisational learning and knowledge management is to ensure that individual learning leads to improvement in organisational knowledge. Successful learning organisations create an organisational environment that integrates organisational learning with knowledge management. Whereas organisational learning is primarily concerned with the continuous generation of new knowledge to add to existing knowledge assets, knowledge management mainly focuses on the capturing, storing, sharing, distributing and coordinating of existing knowledge assets across the organisation, therefore building and exploiting core competencies that yield superior performance. A common feature of both is the sharing of ideas to create and develop new knowledge, enabled by conductive work environment, culture and IT infrastructure.

Meso and Smith (2000) argue that sustainable competitive advantage results from innovation. Innovation in turn results from creation of new knowledge which is substantially dependent on the speed and level of organisational learning. By concentrating on the critical elements of the learning process itself, organisations will be able to achieve sustainable competitive advantage through knowledge-based competencies (Pemberton and Stonehouse 2000). Central to this is the development of cultures, structures, infrastructures and systems which accelerate and sustain the process of organisational learning.

At the same time, developments in communication and information technology have transformed the ability of organisations to acquire, store, manipulate, share and disseminate knowledge, resulting in new management and organisational structure and culture.

## 5.4 Resources, competencies and capabilities

There is abundant literature in the strategic management discipline, including that on the implementation of Strategic Information Systems, that highlights the importance of organisational internal factors in explaining the variation in performance of organisations. This is particularly true when referring to performance over a period of time. In an attempt to explain and measure organisational performance, there is a greater tendency towards viewing organisations from the resourced-based view (see Section 5.6). The resource-based view refers to organisations as "a collection of human and physical resources bound together into an administrative framework, the boundaries of which are determined by the area of administrative coordination and authoritative communication" (Penrose 1959). A resource is defined by Amit and Schoemaker (1993) as any available factor owned by an organisation including IT. Ciborra and Andreu (1998) add that resources are all assets available in the organisation without specific organisational efforts.

In this context, organisational resources do not create value or improve performance on their own. Value is only created by the organisation's ability to efficiently mobilise and use the available resources through effective management. Amit and Schoemaker (1993) define organisational competency as a capacity to deploy resources, usually in combination, using organisational processes, to effect a desired end. The development of competencies requires a number of skills and technologies to be considered simultaneously and in the most efficient way, to perform a particular task. Therefore, it can be stated that competencies reflect the organisation's ability to sustain itself in the marketplace through better utilisation of it available resources.

The deployment of organisational competencies to achieve the organisation's goals is referred to as organisational capability. Teece (2000) defines capability as a set of differentiated skills, complementary assets and routines to provide the basis for firm's capacities in a particular business. Unlike resources, capabilities are based on developing, carrying and exchanging information/knowledge throughout the organisation in order to achieve a particular target. In this context, an organisation's capability can be determined by its future goals as derived from its business strategy. This implies that organisations can use their available competencies, which are derived from the available organisational resources, to plan or establish the required capabilities to meet future targets. Or, a number of required capabilities can be determined by knowing the type of capabilities that the organisation needs to meet future targets. Peppard and Ward (2004) state that an organisation's current capabilities, based on existing competencies, will either be an enabler or inhibitor to the goals that it can achieve. For example, a construction company can implement an e-commerce system in collaboration with its suppliers based on its human and IS/IT capabilities which are ready to support, share and exchange project information online.

Organisations' capabilities can be differentiated by the efficiency of utilising the available competencies and in turn the organisational resources. Competencies underpinning a particular capability are likely to be resourced differently in different organisations and the resources are integrated and coordinated in different ways, depending on the context of each organisation, including its history, people and management (Peppard and Ward 2004). For example, an organisation might have the capability to manage project documents over the

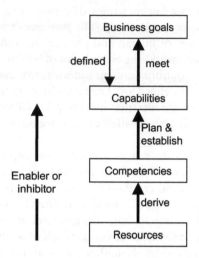

*Figure 5.2* Resource-based models.

Internet through the implementation of Extranets, however the degree of effectiveness and impact on a project's performance can vary according to the type and the way resources are deployed to achieve this goal. An organisation with a strong hierarchical management style, or lengthy decision-making process will struggle more to successfully use an Extranet system compared to an organisation with a flatter management.

### 5.4.1  Core capabilities and learning organisations

It is evident from the above discussion, that a resource-based view of organisations has improved the understanding of the variation in performance of organisations, even those with similar capabilities. Organisations can therefore develop a distinct performance by building up new core capabilities. Core capabilities are defined as those capabilities that differentiate the company strategically, fostering beneficial behaviour not observed in competitive organisations (Leonard-Barton 1992). Thus the capability becomes a core capability when it is:

- valuable – exploits opportunities and/or neutralises threats in an organisation's environment
- rare – the number of organisations that possess a particular capability is less than the number needed to generate perfect competition in an industry
- imperfectly imitable – because of the unique conditions in its acquisition process and/or because of its social complex
- with no strategically equivalent substitutes – with no alternative ways of achieving the same results (Barney 1991).

An organisation can determine a specific core capability to meet its business goal (see Figure 5.2). Such a capability can be created through a transformation process by which an organisation's resources, which are available in open markets to all organisations, are utilised and integrated within the context of that particular organisation. This transformation process makes the "resulting" core capability unique to that organisation and highly dependent on its structure and environment.

Consequently, developing core capabilities is directly related to the organisation's ability to learn, i.e learning how to integrate resources and competencies to deliver and implement such capability. Depending on the current level of "know-how", capabilities can be made more sophisticated by innovative integration of internally available resources (or with external resources) (Ciborra and Andreu 1998). Since learning evolves internally within organisations, core capabilities become highly dependent on the path they took to mature. This feature makes such capabilities difficult to copy and therefore brings about sustainable advantage to organisations.

## 5.5 Managing knowledge: classification of knowledge

In order for organisations to create core capabilities, they need to develop sustained learning environments that facilitate continuous development of competitive skills and technologies. Construction and engineering organisations utilise vast amounts of explicit and tacit knowledge in various areas that are critical to achieving their business goals, e.g. knowledge related to product development and business process integration. Acquiring an ability to manage this knowledge effectively and to create a supporting infrastructure for its use will significantly contribute to the creation of learning environments. This will give organisations a leverage to establish a better collaborative environment within which employees can engage in a continuous learning process.

Successful development of knowledge management (KM)in organisations can face many challenges. Lawton (2001) highlights three main challenges.

- *Technology challenge* Although IT can support KM, it is not always possible to integrate all the different systems and tools to achieve an effective level of knowledge sharing.
- *Organisational challenge* It is a mistake for organisations to focus only on technology and not on methodology. It is easy to fall into the technology trap and devote all resources to technology development, without planning for KM implementation.
- *Individual challenge* Employees, who often do not have time to input or search for knowledge, do not want to give away their knowledge, and do not want to reuse someone else's knowledge.

An analysis of KM failures reveals that many organisations who failed to achieve the intended objective for the implementation of KM did not determine their goals and strategy before implementing KM systems (Rus and Lindvall 2002). In fact, 50 to 60 per cent of KM developments failed because organisations did not have a good KM development methodology or process (Lawton 2001). In many cases, organisations ended up managing documents instead of "meaningful" knowledge simply because commercially available tools, which are marketed as KM tools, are in fact document management tools (Rus and Lindvall 2002).

### 5.5.1 Knowledge classification

In order to successfully manage knowledge in construction and engineering organisations, knowledge needs to be classified according to its processing requirements. The management of such knowledge should not only deal with explicit knowledge, which is generally easier to handle, but also with tacit knowledge. Thus, such knowledge can be classified into the following three categories:

#### Electronic library

The electronic library category contains all the explicit and codified knowledge that is considered valuable to an organisation. Construction and engineering organisations have large amounts of explicit knowledge that needs to be stored, shared, and re-used. Explicit knowledge stored in electronic libraries may include local policies, laws, standards, guidelines, manuals, directories, proposals, contracts, project plans, project-management documents, CAD designs, reports, and information about clients, suppliers and subcontractors.

Information in an electronic library should be labelled and stored in a structured format for easy retrieval. This could be facilitated by the use of the appropriate technologies such as databases, intranets, document management systems, etc.

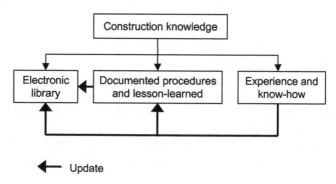

*Figure 5.3* Engineering knowledge interaction.

*Documented procedures and lessons-learned*

Construction and engineering organisations rely heavily on structured and non-structured procedures that produce different deliverables at various stages of the project. Tacit knowledge is normally developed through the experience acquired during the carrying out of this work. This generated knowledge becomes an important internal resource for organisations, which is utilised in future projects. The broad range of relevant knowledge and experiences resulting from such work can be depicted by the following examples:

• Knowledge and insights about procedures, relationships and dependencies that are needed to accomplish certain tasks, such as tasks required to deal with certain problem ands on a particular design or a construction site.
• Amendments to existing procedures as a result of an experience gained from performing a certain task which might have identified the need for modification. This process might also provide tips on how to better perform a particular task or avoid certain mistakes.
• Solutions to problems encountered while performing a particular procedure of, as well as, best practices.

The generated tacit knowledge is embedded in mental models, individual patterns, values, and insights and is extremely difficult to codify, document and transfer to colleagues. Although not all tacit knowledge can be articulated into explicit knowledge that is easier to share, some can. Documented procedures and lessons-learned describe two ways to articulate and internalise tacit knowledge.

DOCUMENTED PROCEDURES

A documented procedure provides professionals with a set of logically ordered activities to reach a goal or accomplish a certain task. Generally, documented procedures provide guidance, suggestions and reference material to facilitate better performance of the intended tasks. They can be presented in a semi-formal computational symbolic notation, i.e. activities and their relationships are presented by formal symbols (boxes and vectors) while additional information can be attached informally. Procedure does not necessarily have to be a "real" captured procedure that has occurred in the past, but it can also be a potential solution of how things could or should be done on site.

Knowledge contained in documented procedures should not be limited to the list of activities that need to be carried out, but should also include the knowledge about why work had or has to be done in that way. Such knowledge can empower professionals to reuse it innovatively in future projects.

LESSONS-LEARNED

Lessons-learned is another form of capturing tacit knowledge and of articulating it into explicit knowledge. This special documentation allows professionals to record lessons-learned form their work experience, share it, and make it available for future use. Lessons-learned documentation covers the full and detailed descriptions of the identification and solutions of clearly explained problems.

The questions raised and discussed during work reflection, which can be documented in lessons-learned, can cover technical issues, organisational aspects or special social situations. Lessons-learned should also include the description of failed approaches and those which are not chosen for implementation.

## Experience and know-how

Experience and know-how refer to the personal tacit knowledge that employees gain from their work experiences and it is hard to verbalise or codify. This tacit knowledge may exist in the form of subjective insights, intuitions and hunches. Design, engineering and construction practices are disciplines that are enforced by accumulation of skills gained through experience, and therefore knowledge in these disciplines is particularly important for organisations to manage. This is critical as it empowers professionals to act more effectively in difficult situations and to plan more efficiently in future activities. Also, knowing what employees know is necessary for organisations to create a strategy for preventing valuable knowledge from being lost.

## 5.5.2 Knowledge transformation

Naturally, the effective use of knowledge in organisations evolves over time. Organisations get more involved in KM as the culture for knowledge sharing is built up, starting from the utilisation of electronic libraries into sharing of tacit knowledge. This process highlights the interdependence between the previously mentioned types of knowledge. This dependence, if recognised and formalised effectively, will ensure continuous updating of an organisation's stored knowledge and therefore lead to better utilisation of knowledge, and thus better learning environments.

To illustrate this point, knowledge stored in documented procedures and lessons-learned could be used to update relevant explicit knowledge in the electronic library. Similarly, tacit knowledge of employees' experiences and know-how could be used to update relevant knowledge in the documented procedures, lessons-learned, and the electronic library (see Figure 5.3). For example, when knowledge gained from an employee's work experience causes the need to amend a certain procedure or modify certain documents, actions should be taken to update the effected knowledge.

## 5.6 Resource-based models

In light of considering knowledge as a key source for innovation and sustainable competitive advantage, and knowledge management as being the tool to achieve that, resource-based models can provide a framework for combining these resources to create core capabilities for organisations. This section introduces two models which enforce the resource-based view of organisations, particularly in achieving the required capability to successfully implement IS/IT projects. They both address effective implementation of IS/IT and the associated organisational learning. The two models help to understand how organisations gain insight into what is involved in creating new competencies and capabilities and set up procedures in recognition of some of the critical factors that need to be dealt with when a new technology is introduced or even before it is introduced.

### 5.6.1 Learning in the capability development process

Ciborra and Andreu (1998) introduce a resource-based model for the learning organisation. The model focuses on the concept of "core capabilities" of organisations. Core capabilities are those that differentiate a company strategically, fostering beneficial behaviours not observed in competitive firms. They portray their model in three main loops (see Figure 5.4):

#### Routinisation learning loop

The first transformation step develops capabilities from standard resources. Two different types of learning takes place at this step: The first aims at mastering the use of standard resources, and produces what is referred to as "efficient work practices". Individuals and groups in the firm learn how to use resources in the context of a given organisational situation. The quest for better work practices may even trigger a search for new resources, more appropriate for the practices under development. Or, the appearance of new resources (say new technologies) may motivate individuals and groups to "take advantage of them" through new work practices. Thus, there is a learning loop between resources and work practices.

#### Capability learning loop

The second type of learning creates capabilities from existing work practices. Several important issues need to be highlighted in this learning loop.

- It relates to work practices to organisational routines.
- The result has a strong potential connotation, as capabilities convey what an organisation is capable of doing if properly triggered, i.e. capabilities

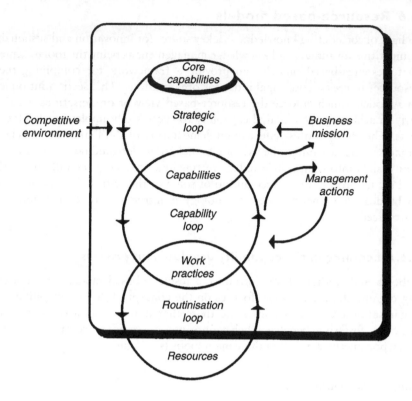

*Figure 5.4* Capability development loops in a learning organisation (Ciborra and Andreu 1998).

involve generalising work practices and putting them in a wider context that defines how they work.

- The result – capabilities – is easily described in terms of what they do and how they do it, but what they do it for is taken for granted, not necessarily well defined and rarely challenged.
- Since the need for new routines for work practices can be detected during the process, it also becomes a learning loop.

### Strategic learning loop

When faced with its competitive environment, a firm learns why some capabilities have strategic potential (they are rare, valuable, etc). A converse influence, from core capabilities to capabilities also exists through the competitive environment, as core capabilities of different firms competing in a given environment (industry) define the "standards of excellence" in that environment, and so they point out what capabilities each firm should develop in order to compete effectively; and it is when confronted with the competitive environ-ment that capabilities acquire a sense of "why" they are important, thus clarify-ing their role and scope.

In addition, capabilities are difficult to imitate in part because of the learning involved in the routinisation and capabilities loops: to develop similar capabil-ities, competitors must go through those learning loops.

An organisation's business mission is also relevant for identifying core cap-abilities. In this context, some capabilities emerge as fundamental to achieving the business mission. Capabilities in this sense are candidates to become core. Again, there is a converse influence. Core capabilities can enable new missions which, if accepted as such, trigger new capabilities, "core capabilities trans-formations". All these interrelationships define another learning loop linking capabilities and core capabilities.

### 5.6.2 Linking IS capability with IS competencies and resources

Peppard and Ward (2004) present a resource-based model to reflect the com-ponents of the IS capability. This model has three levels:

1   *Resource level*  This level refers to the key resources required to develop the IS competencies such as skills, knowledge and behavioural attitudes of both employees and external providers.
2   *Organising level*  This level is concerned with how these resources are

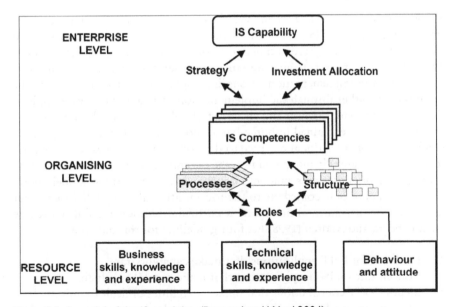

*Figure 5.5* A model of the IS capability (Peppard and Ward 2004).

utilised and managed through the organisational processes and roles to create IS competencies.

As mentioned earlier in this chapter, competencies are embedded in organisational processes and influenced by the organisational structure. The development of a particular competency in an organisation depends on the learning environment of that organisation, i.e. how people apply, integrate, share and disseminate their knowledge. Clearly individuals play a significant role in contributing to the creation of such competencies.

3   *Enterprise level*   At this level the capability becomes visible and ultimately recognised in the performance of the organisation.

Developing capability to differentiate an organisation (core capability) is highly dependent on the organisation's business strategy, particularly if this capability is based on IS/IT. However, implementing such a strategy, i.e. developing the required capability, is dependent on the status of the IT infrastructure, the organisation's ability to deploy the constituent competencies, and to influence the current processes and systems to take the new shape. At the enterprise level, implementing and managing change becomes more critical (Davenport 1998).

## 5.7 Unlocking the business value: embedding IS/IT into core capabilities

While some organisations have managed to gain advantage through successful implementation of IT/IS projects, it is unlikely that they can sustain such an advantage over a long period of time. Technology is no longer proprietary and is available in open markets to all competing organisations. Competitors can catch up through the implementation of similar applications or even overtake the organisation, either through the implementation of a more innovative application, or by deploying a newer and cheaper technology for a similar purpose.

The previously explained resource-based models indicate that IS/IT projects can be a vital organisational resource to develop distinct business core capabilities if they are carefully integrated into the organisation's processes and culture through the organisation's unique competencies. The learning model shows that IS/IT cannot only contribute to the routinisation and capability loops, but it can also be instrumental in making a particular business capability become core. Ciborra and Andreu (1998) list four guidelines for this purpose:

1   Look out for IS/IT applications that make capability rare.
2   Concentrate on IS/IT applications that make capability valuable.
3   Identify IS/IT contributions that make capabilities difficult to imitate.
4   Concentrate on IS/IT applications with no clear strategically equivalent

substitute (functionality of the application cannot be achieved by any other means but IS/IT).

The second model, the IS Capability Model, indicates that sustainable IS capability can only be delivered if competencies are developed across the organisation's departments (and not only within the IS/IT department) through supporting processes, employees' roles and management structure. In this context, Peppard and Ward (2004) state that in order to unlock the capability of IS/IT to contribute to the continuous business improvements, three attributes needs to be considered:

- fusing IS knowledge and business knowledge;
- a flexible approach and reuse of IT infrastructure;
- effective use process to unleash the potential of IT in realising the business value.

In order to successfully implement these attributes, business managers are required to develop IS/IT management skills along with a full understanding of the business strategies (mission, aims and objectives) within organisational context and culture, competitive environments, learning process and behaviour of employees when confronted with change. Such skills will allow managers to recognise and carry out organisational changes that are necessary to create the sought-after core capability which is enabled by IS/IT.

This task is normally delivered and managed differently in different organisations depending on their internal experience and culture. Hence, this enables organisations in developing their distinct sustainable capabilities. Real innovation and continuous improvement, particularly in engineering and construction organisations, cannot be achieved by the implementation of a new technology alone but has to be associated with a constant innovation in products and business practices which are enabled by technology. This requires essential complementary organisational vision and capabilities within which effective innovation can be achieved. This is consistent with the new product development literature which consistently argues that the availability of appropriate organisational capabilities positively affects the outcome of the product development process (Verona 1999).

It is important to note here that the engineering and construction industry is a project-based industry where organisations need to work in a complex supply chain to realise a project. A large amount of project information is generated, amended and exchanged across a large number of organisations during the project life cycle. This process demands that each partner have the ability to capture project information from other partners so that it can successfully execute its role in the supply chain. Developing a core capability in one organisation, to gain a competitive advantage against its competitors, might place that organisation in a difficult position to collaborate with other project partners as it could influence the communication process. Therefore such core capability could have a negative impact on the organisation as a whole. Case study 5.1 explains this issue.

## 5.8  Case study 5.1: Technology-based core capability – strategic failure

A structure engineering design organisation has developed a core capability to produce faster and comprehensive designs for steel and concrete framed building through the utilisation of state of the art technology. Using neutral files to transfer data between applications, i.e. from analysis to design to estimating to manufacturing, project information was seamlessly transferred from one application to another. As a result, this capability has given the organisation a competitive advantage by allowing it to cut the lead time for the design process and to produce comprehensive documentation and quality products to clients. In addition to minimising design errors caused by issues such as information mismatch, lack of specifications, re-keying of project information, etc., the organisation was able to produce steel manufacturing information which can be sent directly to manufacturers, therefore improving and accelerating the manufacturing process.

Although this capability improved the efficiency of the internal processes of the organisation and led to the production of high quality design products with less cost and time, it created a negative impact on the organisation's ability to work with other project partners. Partners such as project managers, architects, quantity surveyors, contractors, suppliers and clients were "expected" to exchange and use project information with this organisation in a similar manner. This caused many problems such as:

1    The organisation produced project information in recently produced standards for data exchange format. This was incompatible with partners" IT applications which mainly used currently adopted "industry standard" format. Thus, partners found it extremely difficult to work with this organisation.
2    Reducing the lead time to produce detailed design products meant that project information could be delivered to other project partners in a shorter time. This in turn will raise the expectation regarding the productivity of other partners particularly by clients.
3    The organisation has constantly pressed clients to take advantage of the capability to produce manufacturing data in order to accelerate the manufacturing process. This has been met with different levels of resistance from manufacturers.

These facts have significantly contributed to the isolation of the organisation in its industry and therefore contributed to its downfall.

# Chapter 6

# Successful implementation of knowledge management

## A holistic approach

## 6.1 Introduction and background

In Chapter 5, the role of "learning organisation", particularly in the field of knowledge management (KM), was clearly highlighted as a key concept which organisations need to adopt if they want to maintain their competitive advantage through IS/IT. The concept of knowledge management has emerged over the past decade to refer to acquiring, storing, structuring and deploying knowledge across the organisation; making knowledge widely accessible through the utilisation of a combination of IT-based solutions and management practices. If effectively implemented, knowledge management can improve performance by decreasing production time and cost as well as increasing quality through avoiding mistakes, reducing reworks and making better decisions. Implementation of this concept can be ensured through a clear implementation strategy, as in any IS/IT project, to increase the likelihood of success and reduce capital investment risk.

The literature identifies two main approaches to KM implementation. The first approach focuses on the capability of information and communication technologies to deliver timely knowledge to employees as and when required. This approach is based on documenting explicit knowledge and externalising tacit knowledge through IT tools. A number of technological tools are available to support storing, refinement, archiving, distribution and sharing of explicit knowledge as well as providing means to transfer tacit into explicit knowledge. The second approach focuses on soft issues such as people and the work environment. This approach considers people as the main element for the successful implementation of KM and thus its key components are improving communication between employees, training and personal development, communities of practice and organisational culture.

The former approach follows the technology route and does not fully address the role of people and the work environment. The latter approach focuses on the soft issues and ignores the importance of advances in technology to provide an effective knowledge-sharing environment. The literature in this field, however, indicates that there is a need to integrate these two approaches to improve their effectiveness, hence leading to the emergence of what is known as the "socio-technical" approach (Koch 2003; Chourides et al. 2003; Shankar et al. 2003; Maier and Remus 2003).

Also, an analysis of unsuccessful implementation of KM reveals that many organisations were unable to achieve the intended objectives of KM systems because they did not determine clear goals for a comprehensive strategy before implementing KM systems (Rus and Lindvall 2002). Lawton (2001) states that 50 to 60 per cent of KM deployment fails because organisations do not have a good KM development methodology or process. In many cases, organisations end up managing documents instead of "meaningful" knowledge simply because commercially available tools which are marketed as KM tools, are, in fact, document management tools (Rus and Lindvall 2002).

An increasing number of organisations have started to realise the importance of deploying a methodology that provides a systematic and specified process for acquiring, storing, organising and communicating engineering knowledge (Price et al. 2000; Nonaka and Takeuchi 1995; Schott et al. 2000; Koch 2002; Sainter et al. 2000; Rus and Lindvall 2002). However, there is a lack of methodologies that fully address KM as an integrated system (socio-technical). In this context, KM implementation methodology might be defined as "processes that facilitate the identification, capturing, development, distribution and effective usage of both tacit and explicit knowledge as an integral part of an organisational infrastructure that allows organisations to effectively manage knowledge in support of their business objectives". Based on this definition and in order for the integrated approach (socio-technical) to be successfully implemented, it should be holistic and should equally address the following:

1   *Knowledge* – to be able to effectively deal with knowledge available in organisations; identifying, capturing, storing and deploying knowledge (knowledge life cycle).
2   *Supporting processes* – to define clear processes that will enable the above tasks to be effectively achieved, i.e. adopting a process-based approach to KM implementation.
3   *Organisational infrastructure* – to establish the necessary supporting work environment to bring the above to fruition.

## 6.2 The need for a holistic KM implementation model

Construction and engineering organisations embrace vast amounts of explicit and tacit knowledge in various areas that are critical to achieve their business goals (Rus and Lindvall 2002; Shankar et al. 2003). Managing this knowledge effectively promises to allow such organisations to save time and money, improve quality and performance and gain a competitive advantage. Despite the growing interest in KM and the number of KM frameworks and methodologies proposed in the literature which tend to emphasise different aspects of KM, there is a lack of commonly agreed ones to guide KM implementation which are holistic in nature and able to fully address the challenges facing KM implementation. These challenges are (Lawton 2001):

- *Technology* – Technology supports KM, but it is not always possible to integrate all the different subsystems and tools to achieve the planned level of knowledge sharing. Furthermore, security is a requirement that the available technology does not often provide satisfactorily.
- *Organisational factors* – It is a mistake for organisations to focus only on technology and not on the implementation methodology. It is easy to fall into the technology trap and devote all resources to technology development, without planning for KM implementation.
- *People* – Employees often do not have time to input or search for knowledge, do not want to give away their knowledge and do not want to reuse someone else's knowledge.

Any developed KM implementation methodology should be holistic and able to assist managers in construction and engineering organisations to achieve the following:

- align the deployment of a KM strategy with the business strategy and to outline the steps of such alignment;
- transfer the organisation's KM strategy to operational level;
- define the stages of KM life cycle;
- identify the various types of knowledge used in such organisations and how they can be managed at each stage of the KM life cycle;
- identify an organisational infrastructure in support of the KM implementation including people, process and technology along with the necessary organisational structure that facilitates the implementation of the knowledge life cycle stages;
- identify the interrelationships between the elements of the organisational infrastructure to improve the efficiency of implementation;
- assess the status of KM in the organisation and determine the areas of weaknesses or "gaps". The route of progress then becomes visible as organisations focus on improving their weaknesses.

## 6.3 Enablers of KM

There is no consensus on the main enablers of KM. Previous studies however have highlighted five factors which are considered to be critical to the successful implementation of KM. These are:

### 6.3.1 Corporate and strategic management

KM is considered a strategic business tool as it can be a key resource for decision-making in the formulation and evaluation of alternative business strategies (Carneiro 2000); sustainable competitive advantage (Meso et al. 2002); customer focus; improving employee relations and development; innovation; and lower costs (Skyrme 1997). These benefits have to be recognised by management and should be incorporated into the organisation's strategic planning.

### 6.3.2 Information technology

Information technology (IT) was initially considered as the central tool for the successful implementation of KM and in the literature on the subject this position still dominates. Survey evidence from KPMG (2000) suggests that in practice many KM programmes are being led from an IT perspective. Davenport and Prusak (2000) suggest that a maximum of one-third of a KM implementation strategy should be devoted to technology with the remaining two-thirds being people-related. Chourides et al. (2003) state that if strategy and people are the principle drivers for KM, then it could be argued that IT is a fundamental enabler. In this context, IT includes technologies such as intranets, group-ware, list servers, knowledge repositories, database management, data warehousing, data mining and knowledge action networks.

### 6.3.3 Human resources

Although some authors believe that information technology is a key driver for knowledge management, others disagree with this view and believe that KM is about people not technology, and that to start from a "computer" perspective would ensure the failure of KM (Davenport and Prusak 1998). Employees are the key source of intellectual capital which is acquired and managed by an organisation's KM system.

Employees propel the organisational learning process. They articulate personal tacit knowledge into the explicit knowledge resident in the organisation's databases, systems and operating technologies. They make personal knowledge available for corporate use. Furthermore, they tap into the corporate pool of explicit knowledge, internalising it into personal tacit knowledge (Nonaka 1991; Davenport and Prusak 1998). Davenport and Prusak (1998) show that employees' productivity, and thus use of knowledge, depends on a complex combination of factors, such as motivation, reward, skill levels, experience, health and even emotional factors.

### 6.3.4 Culture

There are very strong arguments suggesting that successful KM will revolve around creating the right culture and environment (Hibbard and Carrilo 1998). In spite of his long support of human factors as the key to achieving dramatic gains in knowledge development, Davenport (2002) highlights the need for the creation of a supportive environment that will facilitate trust and sharing. It is also recognised that KM refers to changing corporate culture and business procedures to make sharing of information possible (Bhatt 2001). Organisations need to examine social and cultural values, motivation and rewards, trust and willingness to share and individual and team behaviours as a key part of their KM implementation strategy (Scarborough et al. 1999).

### 6.3.5 Organisational structure

Previous studies have shown that traditional hierarchical and bureaucratic organisational structures are heavily reliant on rules and procedures. This hinders the development and transfer of knowledge by preventing risk taking and innovation. These organisational structures have a tendency to reward people on their length of service rather than inventiveness. Furthermore, the different levels and rigid horizontal and vertical divisions in a hierarchical structure hamper the building, diffusion, co-ordination and control of knowledge (Nonaka and Takeuchi 1995).

## 6.4 "SCPTS" three-layer KM model

This section presents an implementation methodology based on a holistic approach to KM (Obaid 2004) which meets the criteria outlined in the previous section. The model integrates the three main components of KM: knowledge, processes and supporting infrastructure. The "SCPTS" model (Strategy – Culture – People – Technology – Structure) is a three-layer model which was developed following an exploratory case-study research methodology (see Figure 6.1).

### Layer 1: knowledge classification

This layer presents the types of knowledge used in organisations and divides it into three categories based on the format and existence of knowledge in organisations:

- *electronic library* containing an organisation's explicit knowledge which can be codified
- *documented procedures and lessons-learned* representing tacit knowledge that has been transferred into explicit knowledge

- *experience and know-how* referring to tacit knowledge that employees gain through their work experiences and that is not easily codified

### Layer 2: KM life cycle

This layer represents the processes required to effectively manage the three types of knowledge, i.e. the components of the first layer. These processes represent the KM life cycle phases:

- knowledge identification
- knowledge acquisition and development
- knowledge distribution
- knowledge measurement and review

### Layer 3: KM facilitators

This layer presents the supporting organisational infrastructure which is necessary to facilitate the implementation of the KM processes. It includes the key enablers:

- strategy
- organisational culture
- people
- technology
- organisational structure

A description of the various layers and elements of the model is presented in the following sections.

### 6.4.1 Layer 1: knowledge classification

Following the widely accepted categorisation of knowledge, the "SCPTS" model adopts the terms explicit knowledge, tacit knowledge and externalised knowledge. The latter refers to tacit knowledge that is captured and transferred into explicit knowledge. The model refers to knowledge in organisations in three different forms: electronic library; documented procedures and lessons-learned; and experience and know-how. (These three forms of knowledge were explained in Chapter 5, Section 5.3.)

### 6.4.2 Layer 2: knowledge life cycle

This layer of the "SCPTS" model is concerned with the identification of knowledge life cycle stages which are necessary to ensure a full utilisation and externalisation of knowledge within the organisation. It provides a "how to

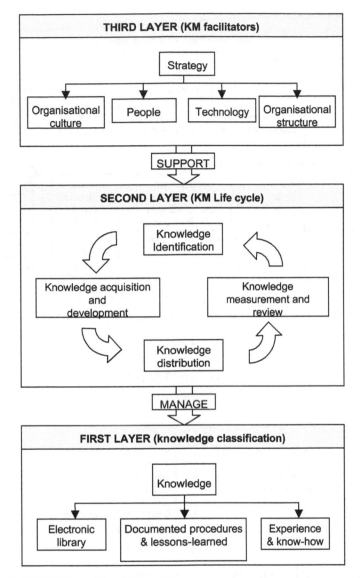

*Figure 6.1* "SCPTS" three layer KM model.

guide" to help managers to understand the flow and relationship between the various stages, and how knowledge (explicit and tacit) is managed at each stage. The model identifies four main stages for KM life cycle: knowledge identification, knowledge acquisition and development, knowledge distribution, and knowledge measurement and review (see Figure 6.2).

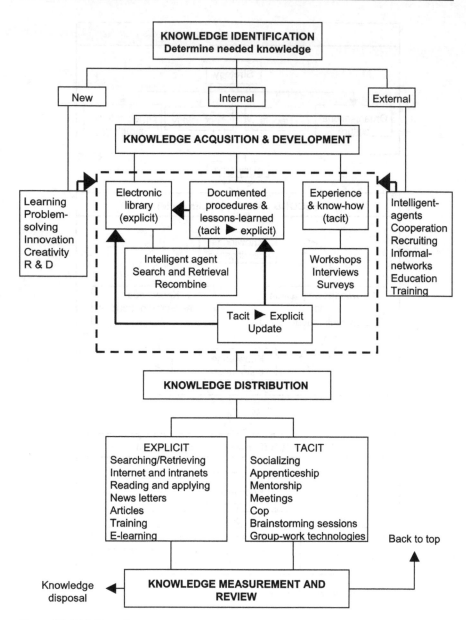

*Figure 6.2* KM life cycle.

*Stage 1: knowledge identification*

Organisations need to identify their knowledge assets as a first step towards acquiring and utilising this knowledge. Management should have the capability to identify knowledge that is considered valuable to the whole organisation, i.e. strategic knowledge. At a departmental level, on the other hand, knowledge should be identified to help departments to better accomplish their tasks or improve their performance.

The knowledge identification process should include all the necessary tasks to raise awareness of the need for creating new knowledge, retrieving existing internal knowledge and acquiring external knowledge. This process should also include tasks that determine the form, convertibility and the ownership of the required knowledge. For example, the following tasks show the logical sequence of accomplishing this process:

- determining the knowledge gap by comparing organisational knowledge needs with the existing knowledge
- identifying the form and convertibility of the required knowledge
- identifying the possible internal and external sources of the required knowledge (internal sources are the knowledge types within the organisation whereas external sources can be partners, suppliers competitors, vendors, etc.)
- identifying the need to create new knowledge

*Stage 2: knowledge acquisition and development*

Having identified the organisation's knowledge requirement, an organisation has to develop plans for acquiring and developing its knowledge needs before distribution. The knowledge acquisition and development stage should include all tasks which are necessary to meet the identified knowledge requirements (creating new knowledge, retrieving internal knowledge and acquiring external knowledge). This stage should also include the manipulation of knowledge (external or internal). For example, converting external knowledge into different types of knowledge for internal use or creating new knowledge through combining and redeveloping internal ones.

Examples of such tasks could include the following:

- creating new knowledge through learning, problem solving, innovation, creativity and R & D;
- acquiring external knowledge through intelligent agents, cooperation with external partners, recruiting knowledgeable employees, informal networks (informal relations with external knowledge sources such as consultants), and employees' training and education;
- retrieving internal explicit knowledge from:

- ○ electronic libraries, documented procedures and lessons-learned through the application and use of technology such as search engines, databases and customised software tools;
- ○ combining and reconfiguring internal explicit knowledge to generate new knowledge: for example, modifying documents stored in an electronic library or using stored lessons-learned to create new knowledge;
- externalising convertible tacit knowledge gained from employees' experiences and know-how to documented procedures and lessons-learned;
- externalising tacit knowledge through workshops, interviews and surveys;
- updating and organising the knowledge contained in electronic libraries, documented procedures, and lessons-learned;
- validating knowledge during development and before distribution: for example a proposed lesson-learned should be subjected to a validation process to verify the accuracy of its content before it is made available for others to view and reuse.

### Stage 3: knowledge distribution

Knowledge needs to be made available and shared throughout the organisation before it can be applied and exploited. Explicit knowledge and externalised tacit knowledge are distributed through activities such as searching and retrieving, Internet, company's intranet, newsletters, articles, training and e-learning. However, the choice of the appropriate method depends on the knowledge complexity level and the nature of the provider and seeker. Easy to capture or internalised knowledge such as laws, local policies, and standards can be distributed on an organisation's intranet bulletin board or through newsletters. Knowledge related to a recipient's field of work and expertise such as new developments in an employee's area of speciality could also be delivered in a simple format such as articles. However, when knowledge is complicated or new to the recipient then training is a necessity, for example training an engineer to install or repair new equipment.

Tacit knowledge, on the other hand, is distributed and shared through formal and informal "socialisation". This takes place in the forms of sharing experiences, spending time with each other, apprenticeship, mentorship, meetings, Communities of Practice (CoP), brainstorming sessions and group-work technologies. Creating the right organisational culture and structure as well as fostering employees' willingness to share their knowledge are essential for sharing tacit knowledge.

### Stage 4: knowledge measurement and review

This stage in the knowledge life cycle is particularly important to contemporary organisations as they operate in highly dynamic technological and global competitive environments. Organisations are considered as distributed knowledge systems, which comprise of knowledge clusters or components (Walsh and Ungson 1991). If these clusters are not reviewed or modified, they usually become passive (Leonard-Barton 1992; Spender 1996). In addition, review of knowledge components is important to deal with environmental stimuli, solve current organisational problems and stress the applicability and risk of knowledge in current circumstances. Review of knowledge is also important because a large part of knowledge, if not used, can be easily forgotten or ignored.

This measurement and review process should include all tasks that aim at justifying and measuring the business value of knowledge, usage and application of knowledge, and reviewing knowledge for updating and disposal. Management has therefore to review and replenish knowledge components continuously in the organisation.

### 6.4.3 Layer 3: KM facilitators

The third layer of the "SCPTS" model constitutes the organisational infrastructure that supports the implementation of the life cycle stages. These are the enablers and drivers for the dynamics of managing knowledge: Strategy, Culture, People, Technology and Organisational Structure.

### Strategy

KM can make a significant difference to improving an organisation's competitive advantage, product or service leadership, operational excellence, customer intimacy, supplier relationship, employee relations and development, and reducing time. If these improvements are linked to an organisation's business goals, then the deployment of KM should begin with a set of goals which are in line with those of the business (see Figure 6.3). A KM strategy then needs to be designed to achieve these goals and should be linked to a top-level measurement system to monitor its implementation and performance. This strategy should aim for a long-term investment in KM to ensure sustainable leverage and reuse of knowledge. It should also strive to identify and "clearly" demarcate the organisational knowledge across various scopes of organisational work practices.

The strategy should have implementation plans to support the acquiring, developing, distributing, measurement and review of the required explicit and tacit knowledge within the organisation. In addition, these plans should involve the creation of an organisational infrastructure that facilitates the implementation of KM life cycle stages and covers the main drivers, namely culture, people, technology and structure. At the operational level, the implementation plans should have clear objectives linked to key performance indicators (KPI) to measure the contribution of KM solutions to the stated objectives. For each measurable objective, an implementation plan must be defined. The deployment of such plans requires the development of operational tactics (see Figure 6.3). The successful implementation of this process will eventually lead to the emergence of new goals at the strategic level and fresh tactics at the operational level that will enable a firm to move up in its knowledge value chain.

### Organisational culture

Organisational culture is critical to promoting learning and development, and sharing of skills, resources and knowledge. The success or failure of an organisation's knowledge life cycle rests heavily on the organisation having an accommodating culture, and its ability to manage and motivate its employees. If organisations do not foster a sharing culture, employees might feel possessive about their knowledge and will not be collaborative.

Many organisations have cultures that do not support KM practices. For example, if employees are accountable for their time and the reward system

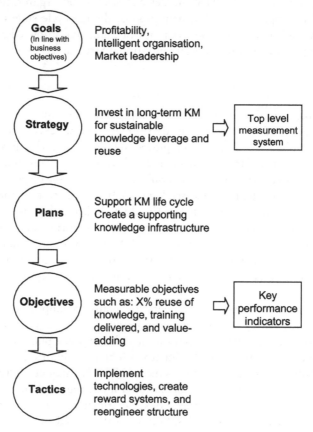

*Figure 6.3* KM strategic planning.

and/or promotions are based on value-added performance, it would be rare to find an employee spending time on knowledge-sharing projects if these are not recognised as value-added activities. Similarly, if there were neither assessment nor credit given for KM activities within the firm, knowledge management would never take priority.

It is evident that organisational culture plays a primary role in facilitating a collaborative work environment. People will not be willing to share their knowledge if there is a lack of trust and respect, or if they sense a lack of interest. A knowledge culture is characterised by the following:

- fostering care and trust among members of the organisation
- seeing failure as an opportunity to learn rather than a punishment
- recording and sharing of knowledge is routine
- visibly rewarding individuals for team work and knowledge sharing
- actively discouraging holding of knowledge and being secretive about best practices
- encouraging asking for help from expert co-workers
- job satisfaction and security
- constantly seeking best practices and reuse of knowledge
- allowing time for creative thinking
- physical space supporting knowledge development and sharing, for example, working in open spaces and providing meeting rooms.

## People

People are the core of knowledge management. This includes employees and managers. Employees are the key source of knowledge owned and managed by an organisation. They are the ones who create, acquire and share knowledge. Managers, on the other hand, have the task of developing knowledgeable employees and creating the environment and infrastructure that is needed to support KM.

People's role in KM is considered in three aspects: managers' role, employees' skills and employees' willingness to share their knowledge. The first two aspects are described below, while the third one is directly linked to the organisational culture as described in the previous section.

### MANAGERS' ROLE

The success of KM requires the involvement of managers at various levels. Top managers should be responsible for providing a KM vision, producing a detailed KM strategy and practising their leadership role. The latter should focus on three aspects. Firstly, it should focus on establishing a culture that respects knowledge, reinforces its sharing, retains its people and builds loyalty to the organisation. Secondly, it should focus on ensuring that middle managers and line supervisors are well informed of the KM strategy and provide them with adequate training, empowerment and support. Thirdly, leadership should focus on establishing an organisational infrastructure that enhances and facilitates the KM life cycle.

In medium and large organisations, there is a need for a knowledge officer, or similar position, at the top management level to coordinate KM activities throughout the organisation.

## EMPLOYEES' SKILLS

Employees must possess the knowledge, skills and experience, and continuously learn and create new knowledge in order for organisations to benefit from knowledge sharing. Therefore, it is particularly important for organisations to develop knowledgeable employees. This can be facilitated through new recruits as well as the training and developing of existing employees. New recruits can be considered as a quick source of acquiring external knowledge. Organisations should also facilitate learning and the acquisition of new knowledge by offering training to individuals in areas where knowledge is needed or desired.

## Technology

Technology is a fundamental enabler to the implementation of KM. Numerous technologies are available, as mentioned earlier, which will not only allow organisations to store, organise and disseminate explicit knowledge but also aid in externalising and socialising tacit knowledge.

Many firms have established knowledge management systems, which have codified knowledge in repositories and linked individuals over the Internet, Intranet and Extranet. Table 6.1 presents some of the available technologies and their roles in facilitating KM processes.

## Organisational structure

Organisational structure can support or hinder the KM life cycle within organisations. It is mainly important for the development, acquisition and the distribution of knowledge. Such a structure should be flexible, flat and decentralised. A flat structure shortens the communication lines between employees and between employees and their line managers. The interchange and development of ideas between specialists in the same field can be facilitated through functional groupings. On the other hand, the use of project teams and groups among functional departments or divisions can enhance knowledge development and sharing.

Cross-functional teams, matrix structures and network organisational structures have proved to be effective in facilitating KM. Additionally, cooperation with external actors like other firms or research institutes, can be a main source of acquiring external knowledge.

Table 6.1 Roles of various technologies in facilitating KM processes

| Technology | Roles in facilitating KM |
|---|---|
| Local and wide area networks | • Collaboration with clients and colleagues: ideally, establishing communities of practice (CoP)<br>• Access to the existing databanks<br>• Access to relevant documents, multimedia files, experts, and training courses<br>• Using modeling and decision support software<br>• Remote access to knowledge bases |
| Databases or knowledge bases supported by search engines | • Develop, acquire and distribute explicit knowledge |
| Document Management Systems | • Develop and organise explicit knowledge, such as project documentation, to be stored and later retrieved for reuse |
| Competence Management (Expert Identification) Systems | • Enables organisations to identify sources of tacit knowledge, experience and know-how of its employees, as a first step of acquiring and distributing this knowledge |
| E-learning | • Utilises computer technologies to create, foster, deliver and facilitate education, training and learning<br>• Provides organisations with practical and cost-effective means of enhancing employees' skills and expertise and distributing explicit knowledge |
| Documented procedures and lessons-learned systems | • Enable organisations to externalise tacit knowledge |
| Artificial intelligence tools based around expert systems and neural networks | • Although still developing, neural networks feature computers that "learn" through experience, thereby mimicking human learning |
| Collaboration tools such as group-ware, video-conferencing, virtual meetings, and CoP | • Enable organisation to share tacit knowledge that is hard to externalise. Such technologies will allow socialising, discussions and exchanging ideas |

## 6.5 Guidelines for implementation

This section briefly lists some guidelines for the implementation of the "SCPTS" KM model.

1   Identify a set of goals that KM aims to achieve for an organisation.
2   Obtain top management support and commitment to KM and prepare for change.
3   Understand the current status of KM in the organisation. This includes assessing the status of the organisational culture, people, technology and organisational structure in facilitating KM as well as the status of knowledge acquisition, development, distribution, measurement and review.
4   Initiate a long-term KM strategy to achieve the identified goals. The KM strategy should:

- aim to identify and demarcate organisational knowledge in various areas
- set KM priorities
- create a KM team and/or identify roles and responsibilities
- raise awareness of KM among employees
- strive to create the required infrastructure to facilitate the acquisition development, distribution, measurement, and review of the needed knowledge
- be associated with a top-level measurement system.

5   Identify the required new, internal and external knowledge. Knowledge identification requires the combination of top-down and bottom-up approaches.
6   Determine whether some organisational issues need re-engineering or improvements according to the organisation's KM needs: for example organisational culture, employees' skills, developing explicit knowledge, distributing tacit knowledge or improving the IT infrastructure.
7   Develop plans and objectives to improve weak areas of KM. These need to be associated with Key Performance Indicators.
8   View progress and adjust as required.

# Chapter 7

# Alignment of strategies and capacity building

## 7.1 Business strategy and strategic planning

The essence of a business strategy (BS) as illustrated by Ramsay (1989: 10) "is to try to arrange things so that you are in control of the situation; ideally, you should break the resistance of competition without a fight. . . . Strategy is usually proactive". The same author suggested that a BS has four main components:

(a) scope of the business – which may be determined by who is being satisfied, what is being satisfied, and how needs are being satisfied;
(b) the resource development or distinctive competence of the business;
(c) competitive advantages of the firm – aspects of the firm's business where it enjoys an edge over its competitors;
(d) synergy – how parts of the firm's processes can best be combined.

The literature on "strategy" ranges from those that focus on excelling in areas of existing processes to those that highlight innovation and/or risk-taking as the drive for a BS. Mulcahy (1990) observes that, to be successful, a construction company must have clear objectives recognising the markets it wishes to address, services it will provide, risks it will carry, structure it will use, the environment it will operate within, controls it will put in place, and returns it wishes to achieve.

To successfully achieve the above, an organisation needs to have an appropriate structure, on-going communication, a team of skilled and motivated people and a culture for performance and satisfaction. Perkowski (1988) suggests that in order for organisations to survive in the face of technological changes, they should:

(a) be eager to understand change
(b) adopt a systems approach to management and budgeting in order to control change
(c) have the pursuit of competitive advantage as the main criterion for any new investments they make

(d)  accept that mistakes happen and reward sensible risk taking
(e)  seek to increase their market share in the long term, endeavour to provide clients with innovative services

The development of a clear and effective BS cannot, on its own, lead to competitive advantage. Efforts must also concentrate on the implementation of such strategies. In this context, the management process for delivering the vision (or mission) of an organisation is known as "strategic planning" (SP). This process is the main vehicle for organisations to channel their corporate energies and resources to address business deficiencies and deliver value-adding benefits to their organisations. Hence, the main aim of SP is to develop and maintain consistency of the implementation activities by defining the strategic approach needed for directing organisations' resources effectively. It requires managers to translate and detail all implementation plans needed to facilitate appropriate course of actions to implement the BS (Venegas and Alarcón 1997). This process needs a series of activities of which the type and nature are dependent on organisational resources and competencies. Thus the appropriateness and effectiveness of SP is highly related to the organisational maturity level and its capability to develop and implement such activities.

Strategic planning should:

1  reflect the constraints imposed by the BS; which includes implementation timeframes, organisational capability, and level of resources available;
2  be directly integrated with the IS strategy (ISS), IT strategy (ITS), and the internal/external business environment.

The conception of the strategic plan should be undertaken at the same time as the development of the BS and, wherever possible, commenced in-line with any strategic change or re-engineering initiatives (Ward and Griffiths 1997). The SP process should be flexible by nature in order to be able to respond to change. Thus, SP can use performance indicators (or goals to measure against) to allow a course of action to be followed. Gaps between the desired levels and the current position are analysed including the evaluation of current performance, availability and suitability of organisational skills, effectiveness of business process and impact of organisational culture on the expected change.

Robson (1997) defines SP as encompassing strategic analysis (mission and goals), strategic choice (options), and strategic implementation (policies, decisions, actions). Figure 7.1 highlights the interrelationships that exist between strategic analysis and strategic implementation.

This model also identifies the following.

(a)  The mission encompasses the main purpose of the organisation, the details of which should be clear and unambiguous. The mission can be used to

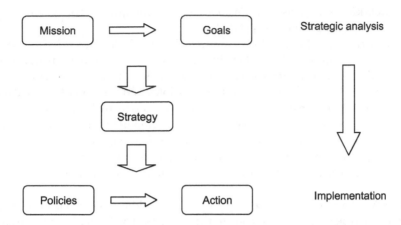

Figure 7.1 Model of the strategic planning process (based on Robson 1997).

define target outcomes, express strong values, and provide guidance for action.

(b) Goals are derived from the mission and are used to define the future position of the organisation (using tangible and measurable objectives).

(c) Strategy provides the means for defining the direction the organisation must follow to meet these goals and achieve the mission. This includes resource issues, capabilities, structure, leadership, culture, etc.

(d) The policy component is used for implementing change issues derived from the strategy (to measure and judge strategy success).

## 7.2 Information system strategies and business competitiveness

Information systems (IS) started to make their impact in the 1950s with the arrival of the business mainframe computers. The format and structure of the IS at that point tended to focus on addressing operational needs. In the 1960s, IS started to address transactional issues to improve overall productivity of organisations. At that time Anthony (1965) developed an IS model that defined the support structure for strategic planning and operational/managerial control.

In the 1970s and early 1980s, the formulation of IS strategies (ISS) started to emerge. The remit of ISS evolved from transactional and support issues, through to decision support methodologies in order to improve business effectiveness. Nolan (1979) developed a six stage model highlighting the stages of IS/IT evolution (as explained in Chapter 3). During the late 1980s (and up to the present date) IS strategy developments have concentrated on providing businesses with a strategic support mechanism, to gain competitive advantage. For the first time, IS started to be considered as a strategic tool to improve business competitiveness and hence started to be integrated into business strategies. The two most well-known models in this field are those developed by Porter and Millar (1985), and Ward and Griffiths (1997).

The Porter and Millar (1985) model (Figure 7.2) focuses on using IS to promote and deliver business objectives. A "Five Forces" model was developed to provide an IS support mechanism that could affect and determine the overall profitability of a business. The principles and concepts behind Porter and Millar's Five Forces model are still often referred to by current research in IS strategy development. However, later developments in IS thinking have shifted towards determining (or assessing) the real "value" of IS to the business. This is especially important, as the costs associated with developing and applying IS strategies to business can often consume a considerable amount of resources.

The Ward and Griffiths (1997) model tries to address the problem of determining the true value of IS systems to the business. The "value" is addressed in four categories; operational; support; strategic and potential to the business (see Figure 7.3).

This model divides the ISS into four key areas (depending upon the classification and perceived value to the business). It allows users to categorise and determine the value of information from a key operational or support perspective, through to strategic and high potential areas. In this context, IS support can be assessed at the same time as the BS is being developed, the process of which enables resources to be prioritised (and aligned) to the most effective process areas to deliver the BS objectives.

To summarise, advancements in ISS development have continued to evolve from the transactional and data processing remits, through to a strategic importance. ISS is therefore seen as a fundamental driver and enabler of business strategy; offering unprecedented levels of business support which can be a unique source of competitive advantage.

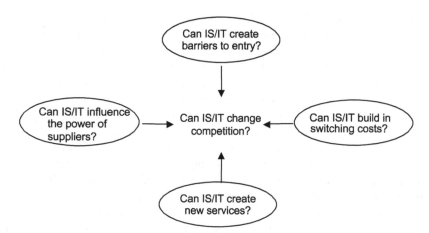

*Figure 7.2* Five forces model (based on Porter and Millar 1985).

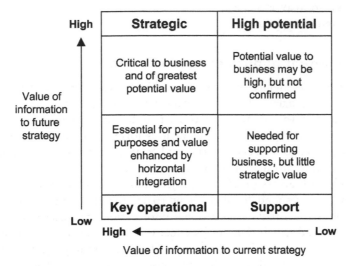

Figure 7.3 Value of information to current strategy (Ward and Griffiths 1997).

## 7.3 Information technology strategies

Information technology strategies (ITS) are defined as highly detailed and structured technological solutions designed to deliver the IS strategy using technology as the primary enabler. The composition and structure of the ITS is therefore predominantly activity based, supply oriented and technology driven.

ITS was first addressed by Leavitt and Whistler (1958). They were the main pioneers in the acknowledgement of a need to develop strategies to manage IT. Their work predicted computer-based systems would centralise information and decision-making power, and have a significant impact on organisational structure and performance. They were the first researchers to identify and describe IT as a combination of computer processing, mathematical programming, operations research and mechanised simulation of thought processes. Later developments in ITS formulation reflected the advancement of technology and its impact on strategic planning (Porter 1985; Hammer and Champy 1993; Ward and Griffiths 1997).

Within construction and engineering, the ITS are used to deliver a range of strategic, operational and support needs, ranging from simple data handling, through to complex decision-making functions. Success of implementation therefore depends on level of access and control of the required information (Abudayyeh and Rasdorf 1991), the result of which can enable IT to be better focused and integrated with the BS needs (Betts 1992) and provide enhanced opportunities for integration (Ahmad et al. 1995). In this context, understanding the exact nature and composition of business and IS needs can affect how the IT strategy is shaped, understood and accepted by the organisation (Mockler and Dologite 1995).

## 7.4 Alignment of business and IS strategies

Construction and engineering projects are highly dependent on gathering, presenting and exchanging information in a useful and logical manner between the various parties involved. Within a project, different types of information are required by various professionals in various formats at various stages of the project. Knowledge workers (a term used to refer to those professionals that exchange project information across the supply chain, as explained in Chapter 1) demand quick access to relevant project and organisational information to enable them to respond to their counterparts in other organisations, while junior professionals require operational information at project level to enable them to undertake their daily tasks effectively. The effectiveness of organisational performance is highly dependent on how successful project information is managed and controlled within and among organisations. This process can only be effectively addressed if a proper alignment between BS, ISS and ITS is achieved. Such an alignment would give the following.

1    A dynamic base of information, which provides

 • a vision for the business information
 • a framework for future communications with other practices
 • a flexible tool for accommodating future business changes

2    Quality assurance; which can be gained by

 • the identification of vital information which might have an impact on business effectiveness
 • providing instant information to daily problems
 • reducing information redundancies and duplication, and process inefficiency through a better understanding of information flow within the business

3    Management oriented value added services, which can improve

 • quality of the decision-making process
 • productivity through effective management

4    Advanced information and communications, which can improve

 • communications within the practice through better understanding of information
 • communications with external information bases (national and international)

5    Progression towards integrated practices, which can be achieved by

- identifying business areas which require information from other practices
- identifying methods for improving communications with other professions in industry

The alignment process can be used to improve organisational efficiency by directly integrating IS/IT with the corporate, strategic and operational needs. This ensures IS/IT resources are "in line" with business imperatives (Galliers and Sutherland 1991). Further enhancement could be achieved by developing the ISS and ITS in unison (to maximise these alignment opportunities). However, this kind of alignment is usually ill considered. Organisations often prepare their BS not knowing the true value and capability of the IS/IT, and how this relationship can be maximised to improve performance (Robson 1997). The BS should be underpinned by ISS which in turn should be supported by ITS. This procedure requires the alignment of needs to meet these demands (see Figure 7.4).

The alignment process identified in Figure 7.4 was extended by Ward and Griffiths (1997) to identify (in more detail) the interrelationships between and needs of BS, ISS and ITS (see Figure 7.5). Figure 7.5 shows the importance of turning ISS to business-based strategies (geared to BS deliverables), and supported by the ITS. The ISS and ITS can be directly determined, from a BS perspective, by identifying the infrastructure and services which are needed to deliver the key business processes.

Organisations should therefore endeavour to assess the capability of their current infrastructure and services to determine the IT and business maxims needed; and either clarify the gaps between what exists and what is desired, or

*Figure 7.4* Alignment of the IS and IT strategies with business strategy.

*Figure 7.5* Business, IS and IT relationship model (Ward and Griffiths 1997).

find a reasonable match between the actual and desired capabilities (Broadbent and Weill 1997). This often requires the adoption of a new management structure or could mean radically changing the organisation's BS, IT infrastructure or arrangements to deliver IS services (Feeny and Willcocks 1998).

The main issue however is for organisations to consider the precise impact of the ISS/ITS on their business objectives and/or their internal operations; particularly, how they can shape their business processes and how they are understood, accepted and used by the organisation (Mockler and Dologite 1995).

## 7.5 Alignment methodology

Based upon traditional information engineering, Alshawi and Aouad (1995) proposed a top-down framework for aligning BS, ISS and ITS. Their model consists of seven independent but related tasks (see Figure 7.6). The framework suggests two levels of analysis; strategic and detailed. The strategic analysis of the business gives the management an opportunity to have, within a relatively short period, a strategic view of their business processes and information before a commitment is made to carry out a detailed analysis. The framework can also assist in prioritising changes according to their contribution to the

*Figure 7.6* Diagrammatic representation of the proposed framework.

achievement of business objectives. Furthermore, business areas (with concentrated business information) can be identified along with their relevance to the overall business objectives. Detailed analysis, on the other hand, can only be applied to high priority business areas which significantly contribute to business objectives.

### 7.5.1 Business and business objectives (strategy)

Establishing a business strategy is a vital prerequisite to a successful IS and IT strategies. It allows management, as well as the implementation team, to focus towards enhancing business activities and to work within clear boundaries. A business strategy should clearly define the primary mission and business objectives of an organisation based on the analysis of its strengths and weaknesses. Questions that need to be answered at this stage are: "Where we are now?", "Where do we want to be?" and "Where could we be?". Business objectives must be clearly defined with no ambiguity and with clear measurable components. For example, a business objective can be:

> To diversify an organisation's services to increase market share, i.e. looking into project management opportunities.

Traditionally, business objectives are vaguely defined with no clear measurable parts which make the interpretation process of these objectives, by the implementation team, very difficult. This could result in equally vaguely focused IS/IT strategies which in turn could become subject to ad hoc changes. This could possibly lead to a major disturbance to planning and implementation of key systems (Clegg et al. 1997).

### 7.5.2 Critical success factors (CSF)

This requires a systematic review of an organisation's objectives, by management, to identify what is critical to achieving them. Emphasis should be placed on establishing measurable factors to enable management to monitor the progress of achieving the identified objectives. Such factors will also provide management with a guide to identify specific information needs and thus highlight critical application areas where business will benefit most. Moreover, CSFs could assist in establishing a portfolio for ISS implementation, i.e. differentiating between systems that require immediate attention and those which can be delayed, based on the business priorities. CSFs should not be more than a reasonable number otherwise the objectives will not be achievable. Examples of CSFs are:

1    Increasing client base by 10 per cent over the next 12 months
2    Reducing services time while maintaining quality

### 7.5.3 Strategic business analysis

The business is first analysed at a strategic level to identify its information needs and processes (work practices). Such analysis must be focused on what the business does rather than on how it is being done, i.e. the analysis must be independent from the firm's internal structure. This allows analysts to examine consistency of business objectives, CSFs and business priorities. Key processes and their key data requirements are identified and evaluated along with their interrelationships. Several techniques can be used at this stage to ensure quality and consistency. One such technique is the data/process matrix. This technique shows which key process uses which key data and whether data are used, created or updated. Business areas (systems) for potential automation can then be highlighted with their relevant implementation priorities (Alter 1990).

It is important that this analysis is carried out first and at a strategic level in order not to get lost in a mass of paper of system analysis. Such processes would require structured and comprehensive methods to ensure consistency and efficiency. Over the last decade, this has proved impossible to carry out manually, especially for large organisations. This has necessitated the development of computerised tools, CASE, to support the work of the implementation team. A CASE (computer aided software engineering) tool is a software environment that assists system analysts and designers in specifying, analysing, designing and maintaining information systems. The main aim of these tools is to improve the productivity and quality of the resulting systems by assisting the developers throughout the different stages of the development process. Upper CASE tools are those which support the upper diagrammatic analysis techniques such as entity–relationship diagrams, data flow diagrams, and structure charts. Lower CASE tools, on the other hand, are those which employ mapping algorithms to transfer formal specification automatically into an executable form. Integrating CASE tools provides facilities throughout the life cycle of an information system, i.e. combined Upper and Lower CASE tools.

### 7.5.4 IS strategy

Based on the strategic business analysis and the CSFs, the implementation team, along with senior management, will be able to identify critical information needs along with their associated systems (application areas). Using clustering techniques of the data/process matrix, potential business areas can be identified. Management's attention must be drawn to these areas to enable the implementation team in establishing an effective and business oriented strategy. Such a process usually produces well-defined and powerful portfolios for IS applications, i.e. priorities of implementing ISs over a defined period of time and in accordance with the firm's available resources. This exercise will allow management to:

1  identify the role of information and its automation in enhancing the business mainstream;
2  evaluate creative ideas about the use of information against business information model;
3  ensure that all investments are driven by the business;
4  evaluate existing systems in a future rather than historical context.

### 7.5.5 Detail business analysis

A detail business analysis is carried out on those areas of the business that have been identified as critical to the business as a whole. The main aim of this step is to establish a conceptual model of a business area which then can be mapped onto a computer level at a later stage. Business processes along with their relationships are identified and analysed at the operational level. Questions like "Where do data originate from?", "Who uses them and why?", "Who manages data and what for?", "Which process is carried out first?", "What technology is currently being used?", "How effective is the current system?", "What improvements are needed?", etc. should be answered. A successful completion of this task requires a structured analysis technique such as data flow diagrams, entity relationship diagrams, and process dependency diagrams. These techniques are necessary at this stage to ensure that the analysis is carried out correctly and that it reflects the real working conditions.

### 7.5.6 Information technology strategy

The IS portfolio established previously should be translated into technology-based solutions, i.e. the technology required to deliver the IS strategy. This implies that IT can only be used effectively for a business if it is utilised as part of the IS strategy that reflects work practices, people and information. Consequently, understanding IT is not equivalent to understanding IS. "Technical staff" usually focuses on the technology, leaving the identification of business needs (process of information) to users. This hampers communication between system developers and end users and is a major factor in the failure of many systems (Edwards et al. 1991).

IT strategy describes the infrastructure and other related services required to satisfy the business needs. Such a strategy requires an understanding of the current use of technology in the business environment, as well as the developments in the technology field and the economics of using it. Moreover, an associated implementation strategy is also required to be developed along side the IT strategy to ensure successful implementation of the new technology.

If efficiently defined, IT strategy can be used to identify the following:

1    opportunities for IT that best meet the business needs
2    existing utilisation of IT resources can be examined and evaluated in terms of its benefits to the business
3    the rate at which new IT applications are adopted, i.e. a rational investment policy
4    the level of impact of IT within firms. This important issue will allow management to balance new IT investments with the necessary organisational change.

## 7.6  Case study 7.1: IS strategy for a quantity surveying practice

This case study illustrates an implementation example of the above framework (see Figure 7.6). It explains the steps taken to establish an IS strategy for a quantity surveying practice. The practice has two offices which are located in the north west of England. It is a typical small–medium quantity surveying practice in the UK with a turnover greater than 1 million pounds. The hierarchical structure of the practice includes partners (executives), senior managers and joiner staff who are responsible for running the business. The practice is engaged with two main surveying functions, quantity surveying and building surveying. The latter is relatively new and is expected to grow in the future.

The study was initially carried out over a period of three weeks whereby all the information and processes were identified. This information was then refined and analysed at a later stage using a CASE tool. Information Engineering Facilities (IEF) CASE tool (Texas Instruments 1990) was used to identify business areas which are suitable for automation and integration. This tool is mainly an Upper-CASE software with powerful and easy to use business analysis functions.

A high level analysis was carried out with the aim of integrating IS/IT with the business objectives. Part of the above stated framework was adopted, i.e. that concerned with the strategic information systems. Factors such as mission, objectives, critical success factors, inhibitors, goals, strategies, and strategic information needs, etc. were discussed with senior management, the findings of which are shown in Table 7.1.

## 7.6.1 Data/process matrix

A strategic information analysis was carried out on the concerned practice. All high level activities performed by the practice were analysed along with their

Table 7.1 Mission, objectives and other business issues of the QS practice

| | |
|---|---|
| Mission | (Indicating its role in society)<br>Is to provide a quality service to customers and to maintain and expand its current profitable level of activities |
| Objectives | (Interpreting its mission)<br>1  An annual growth of 10% in total income<br>2  An annual increase of 20–30% in gross profit<br>3  Diversify the practice's services, i.e. looking for project management opportunities<br>4  Increase quantity surveying activities by 20% over the next three years<br>5  Consolidate reputation: speed up current activities, and improve quality<br>6  Improve the client base: increase the number of clients specially in the new areas |
| Inhibitors | (Things which obstruct the achievement of the business objectives)<br>1.  Insufficient information support and high competition |
| Goals | (To ensure that the business is meeting its objectives at specific time intervals)<br>1  The practice had no specified goals |
| Critical success factors | 1.  Increasing the client base by 10% over the next year<br>2  Reducing service time while maintaining quality<br>3  Obtaining the necessary skills which are required for the diversification<br>4  Improving current staff quality/expertise<br>5  Quicker access to a wider range of information<br><br>(These CSFs are considered to be necessary to achieve the above stated objectives. It was difficult to achieve quantifiable factors due to the lack of objective data regarding the practice current performance. However, all of the above can be measured, over time, against the practice current activities.) |

associated information and information flows. This led to the creation of the CRUD matrix (Create, Read, Update, and Delete) as shown in Figure 7.7. Activities (processes) are shown on the Y axis and their associated data are shown on the X axis. Cost planning function, for example, updates post contract financial report, updates cost plan, reads job plan, etc.

Clustering techniques were performed on the matrix, using IEF CASE Tool. The matrix shown in Figure 7.7 is the output of this process. The purpose of using clustering techniques, as previously mentioned, is to highlight possible business sections that share the same information. From the figure, it can be seen that there are three main business sections. Section (C) represents the creation of the main documents for a project which the business adopts for its daily tasks. Section (A) reflects the majority of the business processes that use (read) the created project's documents. While section (B) shows the office main administration activities.

It has to be noted here that at this high level of analysis, decisions regarding the creation of business sections or information systems are process driven and not data driven. This is because processes defined at such a high level within a practice can be decomposed into their components quite easily by following the practice's procedures. Information and their flows can then be worked out for each system. While with high level data, such as those used in this example, it is almost impossible to decompose them into their components without a clear idea of how and why they are being used.

### 7.6.2 Discussion and recommendations

The analysis revealed that there are three main business sections which need to be examined in further detail. Section (A) in Figure 7.7 represents project costing and monitoring activities which the practice is involved with. Section (C), on the other hand, shows the document generation activities for a project, while section (B) covers all the office administrative work. At this stage of the analysis there is no clear distinction between the information used by both of the surveying parts of the business, i.e. the quantity surveying and the building surveying. However, this picture might change when a detail analysis is carried out on these sections.

As a result of this analysis, three main information systems can be recommended for the practice, a project information system, a project costing and monitoring system and an administration system. It is anticipated that the first two systems will be quit dependent on each other. The project information system will be responsible for generating essential documents that the project costing and monitoring system requires for running its functions. Both of these systems will feed information into the administration system. Figure 7.8 shows the main interaction between the proposed information systems.

Moreover, it is important here to link this information analysis with the previously outlined objectives and CSFs. It is apparent, from the above stated

Cell Values:
- = No reference
C = Create
D = Delete
U = Update
R = Read only

SECTION A

SECTION B

SECTION C

| Business function | POST CONTRACT FINANCIAL REPORT | COST PLAN | JOB DATA | PROJECT PLAN | DESIGN DRAWINGS | SPECIFICATION | CONTRACT DOCUMENT | BILL OF QUANTITIES | CONTRACTOR DATA | TENDER DOCUMENT | CLIENT BRIEF | SITE DATA | COMPANY TARGET | STAFF | CLIENT | PROGRESS REPORT | VARIATION | PRE CONTRACT FINANCIAL REPORT |
|---|---|---|---|---|---|---|---|---|---|---|---|---|---|---|---|---|---|---|
| COST PLANNING | U | U | R | R | R | R | R | R | R | R | R | R | R |  |  | R | R | C |
| PREPARATION OF FINAL ACCOUNT | U | U | R | R | R | R | R | R | R | R | R | R | R | R |  | R | R | R |
| ESTIMATING | U | U | R | R | R | R | R | R | R | R | R | R |  |  |  |  | R |  |
| FILLING AND RECORDS | U | R | R | R | R | R | R | R | R | R | R |  |  | R |  | R | R |  |
| TENDER ANALYSIS | C | R | R | R | R | R | R | R | R | R | R |  |  |  |  | R | R |  |
| VALUATION | U | R | R | R | R | R | R | R | R | R | R |  |  |  |  | R | R |  |
| PROJECT COST CONTROL | U | R | R | R | R | R | R | R | R |  |  |  | R | R |  | R | R | R |
| INTERNAL JOB PROGRESS ANALYSIS | R | R | R | R | R | R | R |  |  |  |  |  | U | U |  | R |  | R |
| JOB SHEET PRODUCTION | U | U |  |  |  |  | R |  |  |  |  |  | U | U |  | R | C | R |
| COMPANY FINANCIAL CONTROL | R | R |  |  |  |  | F |  |  |  |  |  | U | C |  | R |  | R |
| TARGET SETTING | R | R | R |  | R | R | R | R | R | R | R | R | C | R | R | R | R | R |
| ADMINISTRATION SYSTEM | R | R |  |  | R |  |  |  |  |  |  | R | R | R | R | R | R | U |
| COMPANY DATA LIBRARY MAINTENANCE | U |  |  |  |  |  |  |  |  |  |  |  | R | R | R |  | R |  |
| ACCOUNT SYSTEM |  | R |  |  |  |  |  |  |  |  |  | R | U | R |  |  | R |  |
| PERSONNEL |  |  |  |  |  |  |  |  |  |  |  | R | U | U |  |  |  |  |
| TIME SHEET PRODUCTION |  |  |  |  |  |  |  |  |  |  |  |  |  | U | C |  |  |  |
| MARKETING |  |  |  |  |  |  |  |  |  |  |  |  |  |  |  |  |  |  |
| TENDER PRODUCTION |  |  | R | R | R | R | R | R | R | C | R | R | R |  |  | R |  |  |
| BILL PRODUCTION |  |  | R | R | R | R | R | C | R | R | R | R | R |  |  |  |  |  |
| CONTRACT DOCUMENT PRODUCTION |  | R | R | R | R | R | C | R | C | R | R | U | U |  |  |  |  |  |
| SELECTION OF CONTRACTORS |  | R | R | R | R | C | R | R |  |  | R |  |  |  |  |  |  |  |
| SPECIFICATION PRODUCTION |  |  | U |  |  |  |  |  |  |  | R |  |  |  |  |  |  |  |
| DESIGN |  |  | C |  |  |  |  |  |  |  | R |  |  |  |  |  |  |  |

Figure 7.7 Data process matrix, a clustered output.

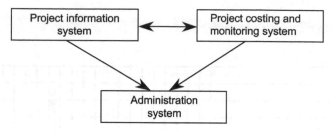

*Figure 7.8* The proposed information systems.

objectives and CSFs, that the practice needs to increase its client base and to diversify into other business areas, i.e. project management. Partners have indicated that this objective is the major objective in the short term (over the next few months). Hence, any investments in IS/IT in the short term should support this objective. This objective can be supported by the proposed administration system if its client's information base is widened. Other objectives and CSFs are based on reducing operational costs to increase profit and to have a quicker access to information which will eventually improve quality of service. Both the project information system and the project costing and monitoring system can achieve these objectives.

## 7.7 IS/IT Implementation issues

Once an IS/IT system has been identified by the management of an organisation, especially of a strategic impact, its successful implementation is a major undertaking. Although each project will have its own best course of action, Paulson (1995) lists a few general guidelines for implementing change into an organisation.

- Maintain openness and honesty throughout the planning, design, development and implementation of an IS/IT process.
- Encourage participatory planning in defining goals and objectives and influencing the design or procurement of the new system.
- Managerial support and involvement should be evident from the beginning to the end of the planning and implementation process.
- The goals for change should be understood and viewed positively by all concerned.
- There should be an effort to coordinate goals of this new system with those of other goals in the organisation, and to maximise overall benefits.
- There must be ample opportunities for education and training on using the new system, and positive incentives for it.
- The organisation and content of the new system must be designed for the people who will use it.

A large number of studies were conducted on the implementation of IS/IT, particularly the effective implementation of Enterprise Resource Planning (ERP) systems in organisations. These studies are mainly related to integrating new processes and structures, as proposed by the new system, into current working practices and how to deal effectively with resistance. Al-Mashari and Zairi (1999) conducted an extensive study on the implementation of ERP and business process re-engineering and have categorised the influencing factors into five groups:

### 7.7.1 Change management

Many authors agree that factors relating to change management systems and culture change management, which involves all human- and social-related changes and cultural adjustment techniques, are the most important factors that are needed by management to facilitate the successful implementation of new IS/IT projects. Revision of reward systems, communication, empowerment, people involvement, training and education, creating a culture for change, and stimulating receptivity of the organisation to change are the most important factors related to change management and culture.

### 7.7.2 Management competency

Good management practices are required to ensure that change efforts are implemented in the most effective manner. The most important managerial practices are top management support and commitment, championship and sponsorship, and effective management of risks.

### 7.7.3 Organisational structure

A new IS/IT creates new business processes which in turn could create new jobs and responsibilities across the organisation. These changes might lead to amending the existing organisational structure. The new structure should determine how the new process should be, how human resources are integrated, and how the new jobs and responsibilities are going to be formalised.

### 7.7.4 Project planning and managements

The implementation process should be subjected to an effective project management which should include factors such as effective planning and techniques, identification of performance measures, allocation of adequate resources, effective use of consultants, building process vision and effective process redesign.

### 7.7.5 IT infrastructure

Factors related to IT infrastructure have been increasingly considered by many researchers and practitioners as a vital component of successful process change and effective implementation (Brancheau et al. 1996; Ross 1998; Broadbent and Weill 1997). Among others, the most important factors are effective alignment of IT infrastructure and BPR, building an effective IT infrastructure, adequate IT infrastructure investment decision, adequate measurement of IT infrastructure effectiveness, increasing IT function competency, and effective use of software tools.

## 7.8  IT training: building the capacity

Successful development and implementation of ISS and ITS in support of business strategies requires high and integrated skills and competencies. Also, IS/IT based capabilities are usually created by the organisation's competencies which are facilitated by the culture and skills of its employees (as explained in Chapter 5). This often requires the provision of advanced and continuous training to acquire the necessary skills needed to deliver these capabilities. These issues are often included in the organisational human resource management (HRM) strategies (Maloney 1997), where the appropriateness of the corporate skill base can be adjusted through the HRM either directly (by recruiting the skills externally) or indirectly (by training the existing personnel to acquire new skills). Managers must therefore appreciate the importance of training, and its relationship with the BS, as "without change in human knowledge, skills and behavior . . . changes in technology, processes and structures are unlikely to yield long-term benefits" (Cooper and Markus 1995: 49).

Training is a management tool and instrument for addressing skill deficiencies. It aims to adapt employee qualifications to job requirements (Krogt and Warmerdam 1997), and can also act as a conduit for linking organisational strategies and goals (Sleezer 1993). This procedure should however be integrated with the long-term needs of the company (Kumaraswamy 1997). In the construction and engineering industry, training has often been linked to improved levels of performance (Naoum and Hackman 1996; Kumaraswamy 1997), but more fundamentally, it can also be used to address critical weaknesses, improve the transfer of skills and knowledge, and help develop a common culture within the organisation. Furthermore, it can facilitate and provide a change in organisational behaviour, which can often enhance an organisation's capability to survive (Kessels and Harrison 1998).

IT training should therefore be integrated with and aligned to the BS needs. This requires the IT training strategy to be structured and tailored to meet specific business objectives, and therefore embraces a number of issues – from a desire to improve overall skill levels across the organisation to increase efficiency, through to gaining of specialist IT skills to attain strategic advantage. In this context, organisations need to spend more resources to train employees in the use of individual IT applications, but they also must develop a longer-term plan to educate users and managers about the capabilities and opportunities that this technology can bring to their business (Breuer and Fischer 1994). It is therefore imperative that organisations reformulate their recruitment policies, performance appraisal procedures and education and training activities to benefit from these IT capabilities (Ahmad et al. 1995). This requires a deep understanding into how IT training can contribute to the successful delivery of the BS needs.

### 7.8.1  IT training strategy and the business strategy

Business strategies try to match opportunities with corporate capabilities (Andrews 1987). Therefore, corporate energies and resources are increasingly being focussed on the concepts and principles of "learning" and the "learning organisation" to help deliver the BS (as explained in Chapter 5). More specifically, as IT has continually demonstrated its role as a core tool and enabler of strategy (Ward and Griffiths 1997; Robson 1997; Rockart et al. 1996), the impact and value of IT training with the concepts of organisational learning promises additional opportunities. However, before the relationship of IT training to the BS can be discussed, it is important to identify some of the main factors that often influence the performance of organisations.

The successful development and implementation of BS can be adversely affected by many factors. Among others, the main three issues are: resources (availability, appropriateness, etc.), organisational culture and skills (matched to the needs of the BS) (see Figure 7.9). Managers should endeavour to acknowledge (and understand) how these issues affect the organisation – to examine the exact role and contribution each of these factors can have on the develop-

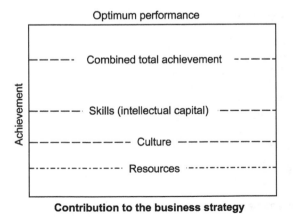

*Figure 7.9* Key issues contributing to the success of the BS.

ment and implementation of the BS, as optimal performance can be influenced by how well these issues are understood. For example:

1   Should resources and corporate energy be spent improving organisational culture and commitment?
2   Or should they be directed towards the provision of training?
3   Perhaps towards buying some new equipment?

Depending upon the organisation, these needs will vary quite considerably, but the overall rationale should aim to prioritise these needs commensurate to their expected contribution to securing BS success.

From a skills perspective, understanding how skills contribute to the successful delivery of the BS is also fundamentally important, as corporate skills are needed to deliver the BS. In this context, training can greatly influence the operation of the core business – often leading to improved business performance (Raghuram 1994; Mata et al. 1995). However, within the construction and engineering industry, advances in IT have continued to place considerable demands on personnel to acquire adequate skills to successfully implement and manage this technology (Heng 1996; Warszawski 1996). However, the needs for such training have not been given the appropriate level of support in organisations (Atkin 1987). This may be due to imposed financial constraints, or could be influenced by the organisation's inability to perceive the criticality of IT training and its impact on the BS (Philip et al. 1995; Ward and Griffiths 1997).

Training should be prioritised to IT tasks deemed critical to the support (and delivery) of the BS objectives. This requires an in-depth understanding of the precise nature of all IT needs required to deliver the BS. For simplicity, these can be categorised into three main areas, specifically, operational, managerial and strategic needs (Goulding and Alshawi 1997; 2002; see Figure 7.10). These needs

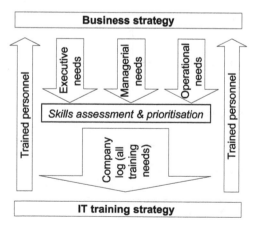

*Figure 7.10* Relationship of the business strategy to the IT training strategy.

should be assessed and prioritised according to their perceived level of importance to the delivery of the BS, the details of which are entered into a company log for subsequent delivery through the IT training strategy (ITTS). At this point, it is necessary to determine the generic IT training needs (common to all users) and specific IT training needs (required by the specialist users).

### 7.8.2 IT training and resource requirements

Organisations should obtain and deploy highly skilled IS/IT resources to meet BS objectives. This process requires the development of skills and competences to meet these needs; which, according to Ward and Griffiths (1997) can be achieved in the following ways:

- training new recruits from school or university
- recruiting experienced staff from other organisations
- training non-IS personnel in application skills
- using external resources (on a long/short-term basis)

These issues are expanded in four key quadrants, as shown in Figure 7.11. These quadrants highlight four areas where resources can be directed to satisfy organisational needs; Key Operational, Support, Strategic and High Potential. Managers can use this matrix to determine the most appropriate use of

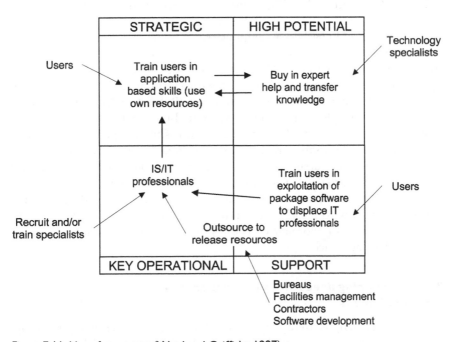

*Figure 7.11* Use of resources (Ward and Griffiths 1997).

resources. For example, one option might be to "buy-in" specialist IT skills (in the high potential quadrant) to transfer this knowledge through training into the strategic quadrant. Other options could include trying to secure long-term capability by moving operational and support resources to the strategic and high potential quadrants.

Thus, training should not be undertaken on an ad hoc or piecemeal basis: it requires a carefully considered and structured approach to map all needs (executive, managerial and operational) into a coherent and flexible training policy linked to key BS deliverables and process enablers. Therefore, the procedure of providing training can often consume a considerable amount of organisational resources; from the outline conceptualisation of the need, through to the efforts extended to designing, implementing and evaluating training outcomes.

Organisations should decide on the most appropriate and effective way of providing training to address deficiencies in their IT skill base. This is normally achieved by one of three options: buying in new skills, training existing staff or a combination of both these options. All have relative merits and demerits, but the amount of resources they consume will depend upon a number of issues, ranging from the availability on in-house training expertise, through to factors associated with the rate of technology transfer, and degree of cultural aversion (Geber 1994; Ingram and Baum 1997; Subramanian and Lacity 1997). In contrast, there is a school of thought that suggests that training should not only be used to address skills gaps, it can also be used to manage social and moral obligations (Scarborough 1997), as the importance of social relations, and the expertise needed to empower the adoption of new concepts, rely heavily upon the distribution of knowledge to employees.

However, most organisations do not have sufficient resources to address all their training needs (Sleezer 1993), the difficulty of which is often exacerbated by the fact that the total costs involved with IT training are very difficult to quantify and almost always underestimated (Robson 1997). This not only includes the initial physical training costs, but also extends to include the time "lost" whilst employees are being trained. If this situation prevails, it is necessary to assess and prioritise training into "essential" needs (core strategic and support needs) and "desirable" needs (those that add benefit, but may not directly contribute to the BS CSFs). It is therefore important to try to determine the exact nature, scope and magnitude of all resources needed, as it often becomes necessary to prioritise these to areas perceived to have the greatest impact on business performance. These priorities can be determined using a needs analysis approach.

### 7.8.3 IT training needs analysis: corporate vs individual needs

The IT needs analysis exercise is a procedure used for determining the exact skills and competencies needed to deliver the BS objectives. This process is undertaken primarily using a skills audit to identify the organisational training requirements, the details of which are then assessed, prioritised, and placed in an organisation log (or skills database). This log acts as a repository of knowledge, and is used specifically to determine the type and level of skills held in the organisation; which can be later matched against the skills needed (the skills gap) to identify the type, level and amount of training required. Furthermore, the skills log can also be linked to the HRM strategy, to ensure that the balance of skills to business need is optimally aligned.

The process of determining the exact nature and scope of IT training needs is however often complicated by the diversity, structure and dynamics of organisations. For example, within the construction and engineering industry, large organisations are often subdivided into smaller groups or subsidiaries (and these may also be further subdivided into sections or divisions). This rationale is often adopted to focus on specialist core operations, or to penetrate niche markets; for example: design and build operations, engineering services, civil engineering, contracting, housing, etc. This arrangement is shown in Figure 7.12. The internal structure of an organisation can therefore make it difficult for assessors to gain an overall holistic understanding of the corporate skills held, especially where a high prevalence of organisational subcultures exist, or where large numbers of employees are present.

The representation in Figure 7.12 distinguishes an organisation containing three discreet groups, each one of which contains a series of sections (or divisions). From the perspective of an IT training needs analysis, the skills audit must therefore include all these areas. However, the dilemma facing managers is that in some cases, groups may be allowed to operate independently from the holding company, which can affect the way in which training resources are allocated. Therefore, the needs assessment exercise is used to categorise the explicit IT skills required at the corporate, group, and section level. This is achieved by identifying the executive, managerial and operational needs in each of these areas, as shown in Figure 7.13. For example, the executive needs

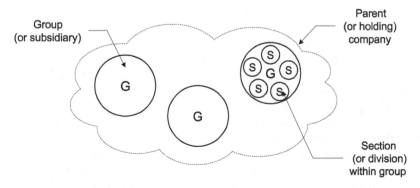

*Figure 7.12* Large construction and engineering organisational infrastructure.

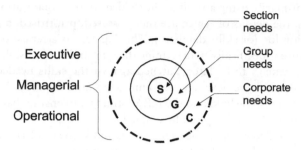

*Figure 7.13* Determination of IT skills – elicitation of needs.

normally embrace the strategic use of IT; whereas the managerial require-ments tend to focus on the coordination and delivery of key objectives; and the operational needs often concentrate on the functional task-dependent IT requirements.

The needs analysis process should therefore reconcile all individual, task and organisational goals (Leat and Lovell 1997; Mata et al. 1995) and balance them against the corporate requirements and personal needs (McCuen 1998; Griggs 1997; Hamel 1991). Awareness of these issues is essential for control, and it is therefore important that all these issues are ameliorated into one coherent ITTS. The composition of the ITTS should include all the IT training needs – from a corporate and group perspective, through to the individual and personal needs.

*Corporate and group needs*

The division of organisations into groups and sections could, in some cases, give these groups and sections complete autonomy from the holding organisa-tion to set their specific business strategies independent of the parent organisa-tion's BS. This provision can allow a greater degree of flexibility to exploit niche markets, without the necessary burden of being tied into the parent organisa-tion's BS and corporate infrastructure. In most cases however, the parent organ-isation will have a corporate BS that will need to be satisfied by the collective actions of the separate groups' activities, as shown in Figure 7.14.

Figure 7.14 identifies three separate groups; Group "A", Group "B" and Group "C". Each of these groups has a separate BS. Their remit and extent is represented by a solid-line circle, and the parent organisation's BS is represented by a dashed-line circle. In this representation, the collective BS's of each of these three groups can be seen to fulfil the parent organisation's BS, and some degree of overlap with strategies exists.

It is also evident from Figure 7.14 that an area of commonality is present in the central core, where all three of the group's BS strategies overlap. In this area, the remits of each of the respective group's BS can be considered similar to each

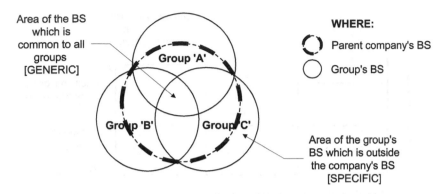

*Figure 7.14* Parent and group BS interrelationship.

other (or identical), and can be thus termed "generic" (common to all). The skills needed in this middle core therefore assume a high degree of commonality exists regarding the skills needed to perform each BS. It also follows that if these skills are similar, then the IT training needs would also be of a very similar nature. On the other hand, the section of a group's BS that does not overlap with any of the other groups' BS, can be said to be unique to that group (and therefore specific). This means that this portion of the BS can be termed "specific", as little (or no) commonality exists in the BS (or subsequent training requirements).

From a corporate ITTS perspective therefore, overlaps in business strategies (with the prevalence of a high degree of commonality), can mean that all duplicated efforts could be rationalised into one corporate generic ITTS (with the specific IT training needs being delivered separately), an example of which is shown in Figure 7.15.

### Executive, managerial and operational needs

Organisations often have to balance a range of skills and competence to meet business needs (K.R. Andrews 1987; Mintzberg and Quinn 1991). These requirements tend to be fulfilled by roles, the denotation and classification of which generally incorporate levels of seniority and responsibility. For simplicity, there are three main role categories used in a business environment, namely executive, managerial and operational.

From an IT training perspective, executive needs often tend to focus on strategic issues aimed at developing understanding and awareness of how IT can bring about competitive advantage to their organisations and how to successfully incorporate it into the organisation's business practices. This type of training usually includes issues such as IT-based core capabilities, IT resource requirements, continuous improvement through IT, IT investments, decision support systems, executive information systems and e-commerce.

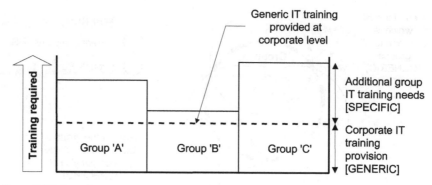

*Figure 7.15* Specific and generic IT training provision.

From a management perspective, the IT training needs also tend to reflect some of the executive needs, but these are often more general in nature, focusing on control-related activities. Typical needs, for example, encompass developments in project management (and planning software), using World Wide Web for business benefits and electronic data interchange (EDI). Operational IT training needs however are often more functional by nature and tend to focus on mainstream business requirements such as professional software packages and exchanging and managing project information.

Whilst there will be some obvious differences in the type and level of IT training required for each of these categories, there is also usually some area of commonality, most notably in the use of mainstream packages such as word processing, the Internet and e-mail applications. It is therefore important to determine the exact nature, level and type of skills needed for each of these categories in the context of the corporate deliverables. These should be assessed, co-ordinated and prioritised according to their level of operational and strategic importance. This process can be achieved using a simple needs audit (as shown in Figure 7.16).

The needs analysis process is normally undertaken using a simple questionnaire, or a series of short tests to determine the prevailing skills levels and corresponding level of training required. Clusters of needs can therefore be readily identified from this matrix, enabling resources to be more effectively targeted to meet the IT skills most in demand (albeit conscious of the BS needs and allocated priorities).

### Individual needs

Individual IT training needs are usually different from the mainstream generic and specific training provision required at corporate and group level. These needs are generally highly personal, and therefore difficult to justify from a business perspective (Krogt and Warmerdam 1997). However, whilst these needs

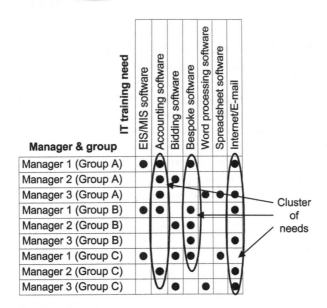

*Figure 7.16* Managerial IT needs matrix.

may not directly benefit the organisation, line managers should contemplate whether these issues could be incorporated into the wider context of the company's HRM development plans, as this approach can often help secure commitment, which could improve employee motivation (Feeny and Willcocks 1998; Neary and Yeomans 1996).

### 7.8.4 Barriers to success

Employees often find it difficult to accept that they do not have the skills required to perform their jobs as a result of the expected change, which can cause organisational friction (Neary and Yeomans 1996). Consequently, the introduction of new technology in the workplace can often cause numerous problems if it is not carefully managed (Bohn 1994), particularly where training is required. Whilst it is generally accepted that employees will often need to be trained (or re-trained) when new IS/IT strategies are deployed, this may also mean that new roles and/or responsibilities are needed (Ross et al. 1996). These issues can tend to cause barriers (and tension), which can have an impact on the overall success of training.

Without change in human knowledge, skills and behaviour, change in technology, processes, and structures are unlikely to yield long-term benefits (Cooper and Markus 1995). Managers should therefore appreciate that users often have a resistance to the implementation of IT in the workplace and that in some cases employees may never fully accept IT (Grindley 1992). Therefore, the decision to implement IT-based solutions into an organisation must evaluate the precise nature of the organisational environment in order to minimise risk (Dhillon and Backhouse 1996). This deliberation should also encompass all human issues and the socio-political factors that could influence success (Powell and Dent-Micallef 1997).

The effectiveness of training can often be influenced by three main factors; the organisation, participants, and the training session. The organisational influence tends to embrace issues of culture, support, commitment, etc. whereas the participants' perspective embraces motivation and relevance; and the training session can influence the type and quality of training provided (which includes the proficiency and competence of the training providers and the teaching and learning delivery methods used). This relationship is shown in Figure 7.17.

### Resource and management barriers

IT training can often consume a considerable amount of organisational resources, from the initial development of the ITTS, through to the costs involved in the subsequent implementation and evaluation process. These costs are almost always underestimated, as managers often fail to appreciate the full resource implications of designing and implementing IT training initiatives (Robson 1997). Resources are not always financial by nature, they can also include other issues such as "time" and corporate "energy". Managers should therefore recognise the myriad of complexities that exist with training initiatives (Miles and Neale 1997), and the subsequent impact they can have on resource allocation.

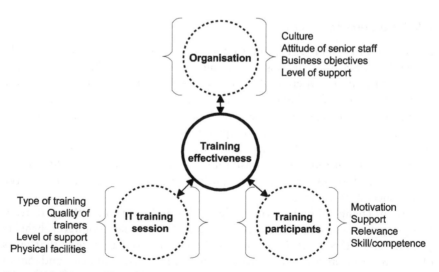

*Figure 7.17* Factors affecting training effectiveness.

On the other hand, the successful design and implementation of any ITTS depends on the organisation's prevailing management awareness and vision. Managers' decisions often directly (or indirectly) affect employees and implementing change. This can cause many difficulties, not because people are resistant to change per se, but because of the way in which this change management process is managed. Furthermore, even the best intentions fail, not because of the ideas themselves, but as a result of the way in which those ideas are shared with individuals (Griggs 1997).

Success therefore, requires managers to fully appreciate "change", the concept of which should embrace the appropriate motivational strategies needed to satisfy the distinct needs of all personnel (M.I. Hall 1997). If managerial attitudes remain defiant and reluctant to embrace change, barriers will develop that could detrimentally affect the success of any organisational training initiative.

### Corporate environment and culture

Culture often encompasses many different notions and meanings, ranging from beliefs, language and ceremonial acts (Meyer 1982), through to perceptions. From an organisational perspective, te concept tends to encompass the internal and external environment, and therefore embodies the whole essence and driving force of an organisation. Corporate culture is often ingrained and very difficult to influence and change. It is affected by employees' deep-rooted values and beliefs, the nature and causes of which can often have far reaching consequences (Mockler and Dologite 1995; Geletkanycz 1997).

The term "culture gap" is the difference between the values and perceptions held by one person, group, or subset; against the perceptions held by others (individuals, groups, or subsets). An example is to consider the views of management against operatives. Each have different perceptions of culture, of each other, but neither are able to identify why this gap exists, how wide it is, or how this gap can be closed. These differences therefore can often instil mistrust and create communication problems, particularly where no common agreement on social rules and protocol exists. Therefore, key cultural barriers must be addressed to maximise shared beliefs and minimise any negative effects associated with misunderstanding. The process of resolving differences requires an in-depth understanding into the root causes of all preconceptions (Mockler and Dologite 1995), which may also require organisational subcultures to be integrated, before any attempt is made to bridge the culture gap (Grindley 1992; Peppard 1995).

From an IT perspective, culture can affect the success of any IT initiative (Davis 1993; Taylor-Cummings 1998), and cultural equilibrium (where all share the same perception) can only be restored when IT becomes part of organisational culture (Earl et al. 1989). This can be achieved through effective management intervention, but this requires acknowledgement that people by their very nature are averse to change, and feel uncomfortable with issues that affect their social standing and working practices (Johnson 1992). Communication and training issues are therefore important to organisational success and whilst managers may not be able to control culture, they can influence its evolution (Walshman 1993). A congruence of shared values should therefore be sought, which links all organisational personnel together – forming a set of shared beliefs, perceptions and goals, directed towards one corporate philosophy and vision (Gunning 1996).

### 7.8.5 Performance gap analysis

From a business performance perspective, understanding how IT skills can contribute to the successful delivery of the BS is fundamentally important. This process should determine what IT skills and capabilities are required to bridge the gap between what is needed, against the present skill and capability base (Van Daal et al. 1998; Goulding and Alshawi 1999). Consequently, the difference between the required performance level and the actual level is termed the "performance gap", as shown in Figure 7.18. (The concept of the performance gap is explained in Chapter 9).

The dotted line indicates the current performance level for area "A", and the dashed line identifies the target level to be achieved after training: both areas shows the performance gap required for the operational, managerial and executive levels. Two performance gaps are identified – specifically, the "opportunity" gap, and the "optimal" gap. The opportunity gap represents the immediate shortfall in skills needed to meet performance expectations, whereas the optimal gap is the highest achievable skill level for that category. Target levels are set by the training manager for the operational, managerial, and executive levels, regarding the perceived impact and import these skills could have on the BS; which in this representation indicates that the executives have the greatest opportunity gap to address.

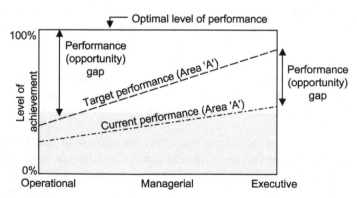

Figure 7.18 Performance gap analysis (Goulding and Alshawi 1999).

Performance analysis can therefore be used to help mangers analyse the scale and importance of the required training. Decisions can be made to rectify performance deficiencies (or further improve skill levels) to maximise perform- ance (Reiblein and Symons 1997). The design, development, and implementa- tion of the ITTS must therefore focus on the planned management of change, using gap analysis to measure performance achievement. In this context, cor- porate skills can be "matched", "measured" and "evaluated" to targets – especially the contribution the ITTS makes to the delivery of BS CSFs.

## Skill and competence issues

The term "competence" can be used to define a particular level of skill, know- ledge or observable characteristic (Prahalad and Hamel 1990), whereas skills are essentially an amalgamation of human expertise and facilities, blended together by the organisation, processes, systems and culture (Klein et al. 1998). Skill and competence levels are therefore crucial for delivering organisational success (Drejer 1996).

From a skills perspective, it is important to understand how skills are nur- tured, developed and influenced. Klein et al. (1998) note that the development of skills was often affected by the organisation, facilities and employees (see Figure 7.19). Deficiencies in any of these three primary areas can have the potential of affecting the development of skills. While the development of competence within organisations can provide access to a variety of markets, contribute significantly to the perceived customer benefits, and be difficult for competitors to imitate (Prahalad and Hamel 1990)

Organisations should aim to build inventories of competence to meet busi- ness requirements, as these can be used to create product advantage (Hamel 1991; Klein et al. 1998), especially when aligned to the BS. They should endeavour to match skill and competence levels to corporate capability, which from an IT skills perspective is fundamentally important (Maloney 1997; Ward and Griffiths 1997).

*Figure 7.19* Corporate Skill Development Influences (Klein et al. 1998).

*Table 7.2* Skills audit questionnaire

| Category | Description (spreadsheets) | 1 | 2 | 3 | 4 |
|----------|---------------------------|---|---|---|---|
| A1 | Sort and categorise cost data | ○ | ○ | ○ | ● |
| A2 | Produce a graph from data | ○ | ○ | ○ | ● |
| A3 | Create external references to files | ○ | ● | ○ | ○ |
| A4 | Produce simple macros for manipulating data | ○ | ● | ○ | ○ |

Note: 1 = no training; 2 = some training; 3 = intermediate training; 4 = full training

Skills and competence levels can be elicited using a skills audit questionnaire, an example of which is given in Table 7.2. In this example, four categories are presented (A1, A2, A3 and A4), and a rating of 1 to 4 can be scored in each of these remits. This type of questionnaire can allow users (or assessors) to determine the type and level of training needed for the skill categories identified. These questionnaires should be completed by all organisational personnel, the results from which can then be collated by the training manager for subsequent analysis.

The skills audit questionnaires can be used to identify training needs and targets more effectively. Moreover, they can be used to highlight specific areas of need from a corporate perspective (where clusters of need are identified). These clusters are however better represented in the form of a histogram, as shown in Figure 7.20. This histogram identifies that the majority of personnel have chosen category "A3" as a particular area of weakness (which indicates that training should be provided). However, if this category was not considered "essential" to the delivery of the BS (or if training resources were limited), then training resources should be diverted to those areas deemed to have a greater impact on the BS (irrespective of the perceived need highlighted by the skill histogram).

In summary, corporate skills should be aligned to meet the specific requirements of the BS (May 1999), and the skills audit can be used to determine the current level of skills (and needs) of the organisation. This type of audit could

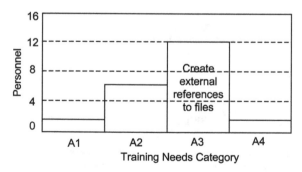

*Figure 7.20* Histogram of skill requirements.

also be used to identify and differentiate between generic and specific IT training needs.

### Generic and specific IT training needs

From an ITTS perspective, the executive, managerial, and operational needs should be addressed using generic and specific IT training. This provision provides an opportunity to help direct resources more effectively (Shirazi et al. 1996). For example, generic training sessions could be used to deliver core IT training issues, such as the use of E-mail and Internet communication, basic software packages, general awareness of IT capabilities, etc., whereas specific training could be used to address specialist needs, which could include EDI, specialist software, etc. In large organisations therefore, this approach could enable generic IT training to be provided by the holding (parent) organisation, and any specific training needs could then be addressed at group level, as shown in Figure 7.21.

The generic and specific IT training arrangement indicated in Figure 7.21, identifies four groups for discussion; Group "A", Group "B", Group "C", and Group "D". These groups all have a particular need for IT training to support their BS, the level of which is identified by the height of the histogram. The shaded area highlights the level of generic training provided by the parent organisation. By default therefore, any training requirements above this line can be classed as a "specific need". Consequently, any group or subsidiary that relied heavily upon IT (or had the majority of its IT training classified as specific), would therefore need a higher proportion of group resources to fund their training needs (as can be seen with group "D").

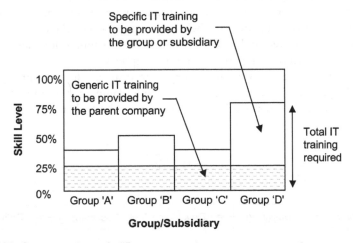

*Figure 7.21* Generic and specific IT training requirements.

### 7.8.6 IT training strategy development

The ITTS should be derived from the organisation's main training strategy. The remit of the ITTS should focus on delivering the appropriate IT skills and competence required to meet the business need. The formation of such strategy must recognise the following.

- Organisations operate in a rapidly changing environment and the ITTS should accommodate such change.
- Knowledge, skills and competence form the basis of organisational intellectual capital (Joia 2000; Mayo 2000), and IT training should be geared to meet current and future IT business imperatives.
- Training needs are often variable, and the ITTS must be capable of being amended to accommodate fluctuations.
- Training (and learning) can be achieved in a variety of different ways, and the effectiveness of each approach should be evaluated in context with the business need.

Developing an organisational ITTS to meet the business need requires six main stages to be undertaken. These range from the initial training definition process, through to the feedback and evaluation stage (see Figure 7.22). This process enables training managers to assess and review their procedural issues regarding the precise role and scope of the ITTS. The first stage (defining) identifies the parameters and issues that need to be addressed (including the metrics required). The second stage (designing) requires managers to design the boundaries of the training initiative (which is often governed by stakeholders, resources, or strict implementation timescales). The next three stages (developing, documenting, delivering) are concerned with determining the most appropriate teaching and learning strategies, documenting decisions and processes, and delivering training in accordance with these requirements. The final stage (feedback) focuses on the evaluation process, where the impact of decisions are measured and assessed against the learning outcomes achieved. Details of this methodology are covered in Chapter 9.

In conclusion, the ITTS should incorporate strategies for measuring change in knowledge, to evaluate the cost-effectiveness and impact of training (Goulding and Alshawi 1997). This requires formal targets to be identified and defined,

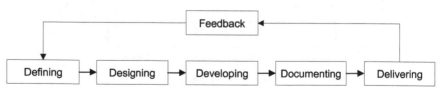

*Figure 7.22* The ITTS development process.

and a demonstrable process (in the form of a training plan) through which this can be achieved.

### Monitoring performance

The allocation of physical and financial resources to IT training often has to be justified to ensure resources are deployed appropriately (Michalski and Cousins 2000). This justification may not always be financial, but the precepts of which should unequivocally prove that benefits have accrued from training. One of the main vehicles by which this can be achieved is by demonstrating improvements in organisational performance. This process should therefore highlight how skills link to (and deliver) the BS CSFs to close the performance gap.

Training evaluation is normally undertaken in two stages, specifically, at an interim stage (whilst training is being conducted) and post-training (after the training has been completed). The interim evaluation stage assesses whether training is being delivered in accordance with the training plan, and covers such issues as progression, appropriateness, and satisfaction with the training providers. The post-training stage however, determines whether the expected skills have been delivered, and the precise contribution they have made to the performance gap.

# Improvement gaps and tools

## Maturity and readiness of organisations

# Chapter 8

# Maturity of organisations and process maturity models

## 8.1 Maturity concept

Maturity usually refers to the adoption of "good practice" within a framework which encourages repeatable outcomes. It is defined by the degree to which organisational processes and activities are executed following principles of good practice. The adaptation of this concept is underlined by a number of assumptions (Becta 2005):

- organisations share common concerns about reliability, sustainability and return on investment;
- a continuous process of review across an organisation will promote continuous improvement;
- while organisations are different, in general their development follows a rational pattern;
- the cumulative effect of a number of small, incremental changes can lead to a qualitative change in how the organisation operates;
- while change continues, there are clear plateaux with well-defined characteristics;
- increased organisational maturity leads to increased organisational effectiveness.

Thus, the basic concept underlying maturity is that mature organisations do things systematically while immature organisations achieve their outcomes as a result of the heroic efforts of individuals using approaches that they create more or less spontaneously. In general, maturity of organisations is related to three main features (Harmon 2004):

1   *Predictability*: the degree of adoption of schedules, milestones, and goals. Immature organisations often create schedules, but then often miss their milestones or goals by wide margins. Mature organisations create schedules and consistently achieve them.
2   *Control*: the consistency within which organisations meet their goals.

Mature organisations always meet their goals with very little deviation. Immature organisations, however, are not sure which goals will be met and have little idea how and when a milestone will be achieved.

3    *Effectiveness*: achieving the right outcomes in an efficient manner. Mature organisations achieve the precise goals they commit to achieve. Immature organisations often achieve some, but not all, of their goals. Moreover, in many cases, the quality may not be as good and the costs may be higher than the organisation would like.

## 8.2 Maturity modelling

Klimko (2001) defines maturity modelling as a generic approach, which describes the development of an entity over time of which "entity" can be anything of interest.

The simplest example of maturity model is the Maslow hierarchy where the entity is the human individual. Maslow (1970) suggested that there is a hierarchy of human needs starting from physiological needs up to self-actualisation needs going through self needs, love needs and esteem needs. During progress through this hierarchy of needs all levels must be achieved to satisfy the human needs.

The entity's development is normally described in a simplified way using a limited number of maturity levels. Each level is described by a set of criteria that characterise an entity at that particular level. Levels are listed in sequential order and the entity progresses from one level to the next one above without skipping any in between.

The concept of "maturity models" was first introduced by IBM in the early 1980s to introduce systematic improvements in organisations' processes. It was noticed that the quality of software developed was positively correlated with the quality of the processes used to develop it. Also, it was noticed that process improvement had to develop in a series of steps, rather than simultaneously. This concept has been widely adopted by organisations worldwide and, therefore, over the past several years a number of "maturity models" have been developed and used by organisations. These models were designed to support improvements in processes, products and delivery. They all aim to help organisations benchmark themselves, to identify the next steps for organisational development, and to illustrate the progression of benefits as an organisation matures.

The main principles of a maturity model are (Klimko 2001):

1    The development of an organisation is simplified and described with a limited number of maturity levels, usually four to six levels.
2    Each level is characterised by certain requirements which an organisation has to achieve on that level.
3    Levels are sequentially ordered, from an initial level up to an ending level (the latter is the level of perfection).

4    During development, an organisation can progress forwards, over time, from one level to the next until it reaches perfection. No level can be left out.

Maturity levels do not provide guidance on how to run an organisation. They are a way to measure how mature an organisation is based on key processes and practices. Therefore, a maturity level is indicative of the effectiveness and efficiency of that organisation and the probable quality of its outcomes.

## 8.3 Maturity models and organisational capabilities

The resource-based models discussed in Chapter 5 indicate that in order to develop an IS/IT capability in an organisation, that organisation not only needs to make the necessary basic resources available but also to have the competencies to bind them together in the most efficient way. Peppard and Ward's IS Capability Model (Figure 5.5) shows that the right resources need to be translated into roles and responsibilities in all concerned departments, i.e. those that are affected by the implementation of the IS/IT system. New roles and responsibilities should empower employees to challenge their work practices and the impact they have on their current organisational structure and processes. If successfully undertaken, these changes can bring about new competencies to organisations. However, this can only be realised if they are supported by the right investment and are considered in the business strategy of the organisation.

The resource-based models can be utilised to explain the concept of progressive achievements of the maturity levels. Maturity models state that in order for an organisation to migrate from one maturity level to another it has to develop the necessary capabilities required by that particular level. According the resource-based models, the development of an organisation's capabilities is highly dependent on the organisation's culture and management structure. This explains the difference in ability among organisations to develop IS/IT capabilities and hence the importance of adopting the concept of "learning organisations" as is explained in Chapter 5. Therefore, organisations that have the ability to develop and successfully implement new capabilities will embrace the requirements of a particular maturity model more effectively and efficiently than those organisations which lack the ability to adapt to the new changes.

Similarly, product-based maturity models (e.g. Nolen, Earl, etc., see Chapter 3), show that organisations can only reach the level of strategic IT applications when they have gone through the various stages of growth. The speed and efficiency of such growth again depends on the organisation's culture and management structure. Subsequently, the time and cost required to migrate from one maturity level to another will vary from one organisation to another.

These issues raise an important challenge. Is there a "fast-track" for a low level maturity organisation to reach a higher maturity level? This challenge becomes a key one if an organisation wishes to develop a core capability (associated with a high maturity level) in order to improve its performance and market share through competitive advantage.

In order to address this challenge, two highly related issues must be considered: time and resources which are required to develop organisational competencies. Time is of essence to organisations particularly in competitive markets. It is just not possible for an organisation to wait a long time to develop the necessary competencies. However, these competencies cannot be effectively developed without first creating the organisational learning environment and the right work environment for them to be developed.

The time required to develop a particular capability is highly dependent on the organisation's current ability in pulling together the basic resources and transferring them into an organisational capability. An organisation with flexible competencies can quickly and easily go through the progression ladder of a maturity model compared to an organisation with limited competencies.

For an organisation to develop an IS/IT capability, it needs to evaluate its current maturity level in terms of processes, structure and work environment so that effective decisions can be made towards achieving such capability. In addition, developing an IS/IT capability to successfully implement IS/IT projects does not only depend on employees at the IT department/unit but also it has to reach out to all the concerned employees in other departments. For this reason, managers need to have a "forward looking" management tool which will enable them to

1  measure their current capabilities in the relevant areas, i.e. those that affect the development of the required IS/IT capabilities
2  predict the required level of change and the associated resources to develop the target capabilities.

Such a management tool should enable managers to address the above issues by identifying the organisation's "readiness gap" for developing and adopting specific IS/IT capabilities. Also, it should provide them with a vision of how best the required capabilities can be developed so that they are able to better plan and utilise their available resources and competencies. Chapter 9 explains the "readiness gap" in detail.

## 8.4 Process improvement maturity model: the case of construction

Information technology has found an increasing role in improving business practices of organisations, from improving the management of supply chain to delivering high quality products and services with greater speed and less cost. Highlights of such improvements can best be planned utilising the process approach to business improvement. The latter has been increasingly favoured in investigating business efficiency and effectiveness as it provides descriptions of work activities and places little emphasis on the vertical function boundaries of organisations, i.e. the typical functional approach to managing organisations. Reports like Egan (1998) stress the importance of the process approach for improving the productivity of the construction industry.

### 8.4.1 Process capability and maturity

An organisation's ability to devise and enforce processes is not the same as an ability to perform them (Construct-IT 2000). Process capability enables organisations to predict the outcomes of a process before it commences, i.e. it is forward looking. In contrast, process performance assesses the actual results of a business process after the process has been completed and thus provides historic data on the project, i.e. it is backward looking. Therefore, process capability focuses on expected results and makes the outcome of the project more predictable. This is an important issue for clients as well as for the construction organisations. As construction projects are often unique and can lead to many challenges, an organisation with process capability can accurately predict the expected outcome of these projects.

On the other hand, process maturity is the extent to which an organisation is able to define, manage, measure and control its business processes. Having process maturity implies that an organisation has the potential to improve its process capability and suggests that business processes can be applied consistently in projects.

### 8.4.2 Standardised Process Improvement for Construction Enterprises (SPICE)

SPICE is a systematic step by step process improvement framework for the construction industry which was developed at the University of Salford (Construct IT 2000). It specifically draws on the capability maturity model (CMM) (see Section 8.6 of this chapter) and borrows many of its basic concepts and adopts them into a construction specific framework.

The SPICE framework supports continuous process improvement based on many small and evolutionary steps (Sarshar et al. 2000). It defines these steps in five maturity levels, which lay successive foundations for continuous process improvement. These maturity levels form a scale for measuring the capability of a construction organisation's individual processes and its overall process capability. They also provide guidelines on how to prioritise efforts at process improvement.

Each level of maturity consists of a set of key processes (see Table 8.1). An organisation can only be at one level of the model at any one time when it satisfies that level's key processes. If an organisation is at Level 1, but implements some of the key processes of Level 3 or 4, it is still considered a Level 1

*Table 8.1* SPICE maturity levels and key processes

| Level 5 Continuously improving | Key processes from original<br>• Process change management<br>• Technology change management<br>• Defect prevention |
|---|---|
| Level 4 Quantitatively controlled | Key processes from original<br>• Quality management<br>• Quantitative process management |
| Level 3 Well defined | SPICE recommended key processes<br>• Organisation process definition<br>• Organisation process focus<br>• Integrated design and construction management<br>• Construction life cycle engineering<br>• Training programme<br>• Peer reviews |
| Level 2 Planned and tracked | SPICE key processes<br>• Brief and scope of work management<br>• Project planning<br>• Project tracking and monitoring<br>• Sub-contract management<br>• Project change management<br>• Health and safety management<br>• Risk management<br>• Project team co-ordination |
| Level 1 Initial | |

Source: Construct-IT 2000

organisation. This is because each level lays successive foundations for the next. By following these levels, an organisation can achieve effective and continuous improvement based on incremental steps.

Most of the efforts of the SPICE project concentrate on defining and raising confidence at Level 2 of the model, i.e. "Planned and tracked". As shown in Table 8.1 eight key processes are identified:

1  brief and scope of work management
2  project planning
3  project tracking and monitoring
4  subcontract management
5  project change management
6  health and safety management
7  risk management
8  project team co-ordination

*Process enablers*

In order to implement the SPICE framework, "process enablers" were established. These are activities that provide detailed features of the key processes that an organisation must possess in order to yield successful results. They focus on the expected results from a key process which indicates an organisation has process capability before a process takes place, i.e. adopting a forward-looking approach. These process enablers are:

*Commitment:* the organisation takes action to ensure the process is established and is lasting. Commitment ensures that leadership positions are created and filled, and that the relevant organisational policy statements exist.

*Ability:* describes the conditions that must exist before a process can be implemented competently, i.e. resources, organisational structure and training.

*Verification:* checking that activities are performed in compliance with the agreed process. The focus is on external verification of processes.
*Evaluation:* involving basic internal process evaluation and reviews. The focus here is on the project team's internal improvement efforts.

*Activities:* describes the activities, roles and procedures necessary to implement processes, i.e. planning, performing and tracking.

## 8.5 Case study 8.1: a client–contractor collaboration

This case study is concerned with a "Pavement Team" which is a long-term partnering project that maintains and develops aircraft taxi and runway infrastructure at an airport. The contract has an annual turnover of approximately £20 million. Due to the nature of the contract, the project has ambitious goals to reduce costs and delivery times whilst still maintaining or improving quality. Prior to the SPICE study, the project was already committed to continuous improvement in an attempt to achieve these goals, although such initiatives focused specifically upon technical processes.

### 8.5.1 SPICE assessment

Two SPICE assessments were performed, one with the project management and design team who were based off-site, the other with the construction management team who were based adjacent to the construction works. The assessment involved the participation of ten and eight members of staff respectively, selected from a vertical section of the organisation ranging from senior management to agents, ensuring an unbiased view of the organisation's performance. The

SPICE performance benchmarks represented Figure 8.1 summarise the assessment findings.

The SPICE assessment noted the Pavement Team's significant steps towards the adoption of manufacturing philosophies as part of their technical process improvement initiative. Moreover, the close physical integration of the client, design and project management teams – the result of the partnering arrangement built into the contract – was highlighted as a major strength. However, the assessment also highlighted several important deficiencies in the project's management processes. The partnering companies' processes and systems were poorly integrated, resulting in duplication of effort and poor coordination. Significant cultural issues were also highlighted, such as the fragmented project and site management teams, which restricted effective communication.

### 8.5.2 Recommendations

A workshop was carried out to discuss and generate improvement suggestions to address these and other issues highlighted in the assessment findings. The SPICE assessment culminated in a number of recommendations for consideration by the project team. Here are some examples of the recommendations.

- The systems and processes of the client and the major contractor are not aligned and understood by both parties. In particular the client adopted a process protocol which does not fit into the contractor's processes. The partnering companies need to integrate processes and systems to develop a bespoke Pavement Team set of processes that reflect the unique way in which the contract operates and which improve communication between the project and site teams.
- Develop a more rigorous induction programme for new staff. This is especially important due to the inevitable turnover of staff on a contract of this length and the project's unique operations.
- The risk management processes need much improvement. Increase the participation of the site management team in risk management to raise awareness of risk mitigation activities at all levels of the project team.
- Undertake a periodic review of the project's processes and systems to develop a culture of continuous improvement.
- The client was at Level 1 of the model. Once the above recommendations are implemented, the client will become Level 2.

### 8.5.3 Benefits

The project's senior management team feels the SPICE assessment provides them with increased visibility into the Pavement Team's performance, and provides a valuable aid in bridging the communication gap between senior management and construction practitioners at a site level.

The project director at the Pavement Team emphasises the following key benefits from using the SPICE model:

- highlights specific improvement areas they were previously unaware of
- provides a base line for initiating improvements at a project level
- demonstrates to the client transparent process improvement over time

The SPICE assessment highlighted significant opportunities for improvement away from the technical issues the team had traditionally focused upon. Consequently, the project team is now planning to widen its process improvement initiative to include management processes.

## 8.6 Maturity models: the case of e-government

E-Government is gaining popularity worldwide. In the long run, the eventual success or failure of e-government will depend on the value it adds to citizens' lives, its government services as well as the cost savings that can be achieved. Some of the values e-government can add are:

1  24 hour and seven day accessibility
2  active citizen participation (Wimmer 2002)
3  open government (an essential component of e-democracy)
4  public access to information (Doty and Erdelez 2002)
5  avoidance of physical trips to government offices
6  avoidance on the part of the government to maintain brick and mortar type facilities to handle citizen services (Kaylor et al. 2001)

Literature is replete with success factors for implementing successful e-government systems. Some of these success factors are:

1  ensuring ability to use required technologies (Borins 2002)
2  educating citizens about the value of e-government (Jaeger and Thompson 2003)
3  developing methods and performance indicators

There are several challenges associated with e-government as well. The General Accounting Office (2001) report identifies these challenges as:

1  sustaining committed executive leadership
2  building effective e-government business cases
3  maintaining a citizen focus
4  protecting personal privacy
5  implementing appropriate security controls
6  maintaining electronic records

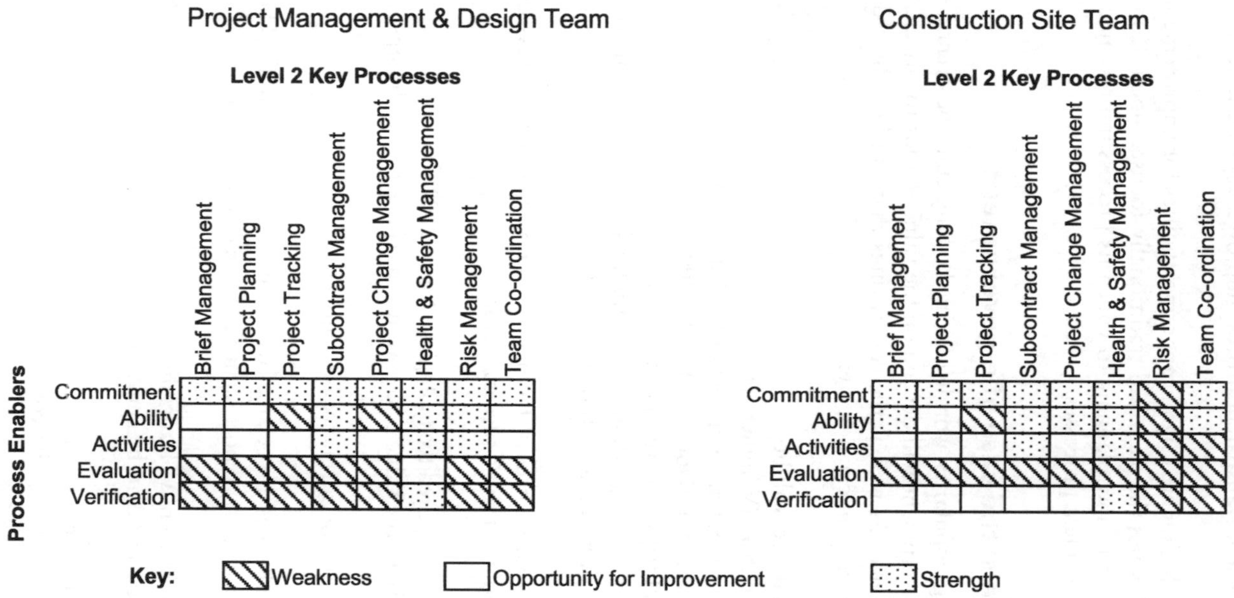

Figure 8.1 SPICE performance benchmark.

7  maintaining a robust technical infrastructure
8  addressing IT human capital concerns
9  ensuring uniform service to the public

The success of e-government projects, i.e. service transformation and IT management in government organisations, can only be monitored, evaluated and confirmed if their stated goals and objectives are measured. Hence, performance measures should be defined for e-government projects to assess their progress and success, giving the public assurance that these systems are accountable to their customers (citizens, industry and other government organisations). Such measures should be developed with stakeholder input and should be clearly documented.

The concepts of maturity models were adopted in e-government initiatives in order to:

1  assess the maturity of processes within a government organisation
2  determine organisational readiness to deliver on an e-government agenda
3  guide the organisation in selection of appropriate process improvements (M. Taylor 2003)

Layne and Lee (2001) present a four level model for the implementation of e-government (see Figure 8.2).

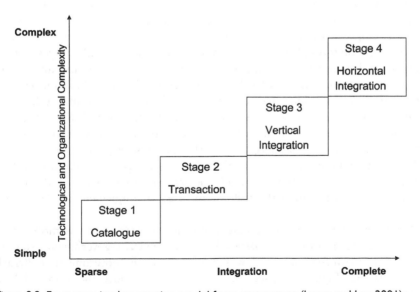

*Figure 8.2* Four stage implementation model for e-government (Layne and Lee 2001).

The model proposes four sequential stages (levels) which a government organisation might progress through to realise the full potential of e-government. The initial level is the cataloguing level where government organisations upload information and limited functionalities such as government forms and make them accessible to citizens through the Internet. The second level is the transaction level where transactional capabilities are added to the website to allow a customer to submit a form, a request, or make a payment online. The vertical integration level involves the complete integration of back office processes with the front end ones. The horizontal integration level is concerned with the integration across government organisations.

In the UK, an e-government model was developed with a focus on reviewing the capacity and ability of IT to support the UK's electronic economy. The model uses five stages to depict the maturity stages for delivering e-services. The lowest stage illustrates organisations that do not have a clear vision, plans and measurement process for delivering e-services. The highest stage depicts organisations that are mature and have an integrated, proactive approach towards delivering e-services. This model was adopted by other countries and used to compare a particular country's service with other leading authorities.

All of these models are flexible enough to be implemented in most of the situations. However, as Layne and Lee (2001) point out, most of the e-government initiatives are chaotic and get unmanageable very soon. It seems that government institutions worldwide are far from maturity, cost savings, revenue generating and downsizing, in contrast to what the rhetoric suggests (Moon 2002). So far, it has led to mild encouragement at best regarding the internet technologies to reinvigorate local governance (Musso et al. 2000). One of the major reasons for this drawback has been the technology-driven implementation methodologies for e-government. Typically, Stage 1 is a simple website; stage two builds up on the first stage and incorporates a two-way communication through electronic data interchange and e-mails. Stages 3 and 4 involve implementing an enterprise-wide system like an ERP and integrating the current system with the ERP. For the Hiller and Balenger (2001) model the fifth stage involves adding technical features like encryption and authentication, chat rooms, more sophisticated interfaces and a seamlessly integrated system. As is apparent from this description, these models are technology driven, and move from one stage to the other progressively incorporating more advanced technology. The success of an IT project, as explained in Part 1, is driven more by the attention paid to organisational issues and process improvement, rather than the technology used to implement it (Arif et al. 2006).

However, a number of e-government models have started to give more attention to the organisational soft issues. Based on the UK e-government five-level model, an e-government model was presented by the government of British Columbia (M. Taylor 2003; see Figure 8.3). In this model, the degree of ICT maturity consists of four measurable components.

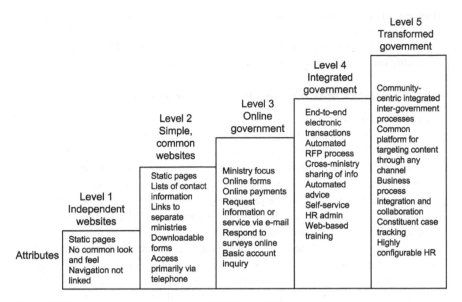

Figure 8.3 E-government maturity model (M. Taylor 2003).

*Environment* measures the extent of the existing supporting activities to the online services, e.g. a clear e-service vision, politician support and legal and regulatory frameworks.

*Readiness* measures how "ready" citizens and businesses are to adopt and use e-services that are available, e.g. do citizens have the required skills? And can they easily access technologies that support e-services?

*Uptake* measures the usage and level of adoption of e-service, e.g., determining how broadly an e-service is adopted by citizens.

*Impact* measures the impact that occurs within the different stakeholder groups as a result of using e-services, e.g. allowing citizens to avoid waiting in lines to obtain licenses.

## 8.7 Other maturity models: the case of software development

This section briefly covers other maturity models, which exist intensely in the literature, in the field of software development. In contrast to the product-based measurement models, explained in Chapter 3, these models are process-based maturity models that are developed as IS/IT measurement tools.

A number of process-based maturity models were developed in the software industry with the aim of improving the final developed product. Some of these models are: Capability Maturity Model, People Capability Maturity Model

(People-CMM), Personal Software Process, Team Software Process, Structured Process Improvement for Construction Enterprises, Trillium, Bootstrap, Software Process Improvement and Capability dEtermination (SPICE), and TickIT. The following sections briefly explain each of these models.

### 8.7.1 Capability Maturity Model: the staged view

The most well-known method for assessment of the software development processes is the Capability Maturity Model (CMM) (Paulk et al. 1993). Capability models began with SW-CMM which was created by the Software Engineering Institute (SEI) of Carnegie Mellon University in response to problems in the software industry such as late, poor-quality, and overrun software projects (Gibbs 1994). The model was requested by the US Department of Defense so that it could characterise software organisations with the potential to be a contractor to the DoD's software products.

The SW-CMM established a five level ladder of process maturity for organisations to climb (see Figure 8.4). An organisation improves several key aspects of its processes – called key process areas (KPAs) – at a time, in order to achieve a new maturity level, and then improves additional areas required for the next level. A KPA has one or more goals describing what should be achieved in that KPA in order for an organisation to be elevated to the next level of maturity. The model also provides the activities that help organisations to implement the goals for each of the KPAs' requirements.

### 8.7.2 Continuous Capability Model (SE-CMM): the continuous view

In December 1994 a new group of eight organisations, in collaboration with SEI Carnegie Mellon University, adapted the SW-CMM version and released a new version known as "SE-CMM". The EPIC group (Enterprise Process Improvement Collaboration) then released the approach's appraisal method in the following year, 1995.

This model uses an architecture called the "continuous" view, derived from a draft of the international Software Process Improvement Capability dEtermination (SPICE) project guidelines (see El Emam et al. 1998). The model is divided into areas for improvement called process areas. A process content is evaluated by assessing compliance with a set of base practices that constitute a process area. Process maturity is evaluated by applying a generic scale of levels to each process area separately.

Each subsequent level is achieved by incorporating the process maturity elements (generic practices) associated with that level. Essentially, generic practices address the manner in which the process content is performed. For example, eventually there is a plan and a process for defining operational concepts, and people are trained to define operational concepts in the manner that the organisation has measured to be most effective.

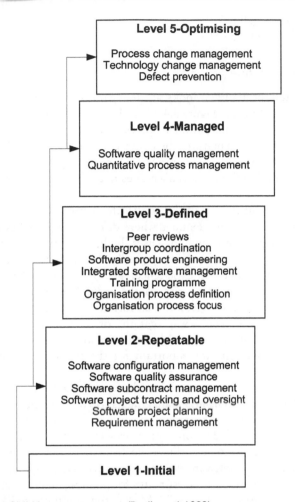

*Figure 8.4* SW-CMM key process areas (Paulk et al. 1993).

The model allows the organisation to determine the maturity of each process area separately and does not prescribe an order in which the process areas should be implemented. The continuous-view consists mainly of a generic non-industry-specific measurement (generic practices), which makes it suitable for use as an integrated part of a larger "general" model that can assess the overall organisational maturity.

### 8.7.3 Capability Maturity Model Integration (CMMI)

SEI also introduced the CMMI which integrates the staged-view SW-CMM and the continuous-view SE-CMM. This section briefly shows how the staged- and continuous-view architectures are implemented, and highlights similarities in the concepts behind the levels.

### Staged-view architecture

SW-CMM (Paulk et al. 1993) looks at an organisation's maturity in a single collective view, as shown in Figure 8.4. Compliance with a KPA in a staged-view model is binary (the organisation's processes comply or they do not). The organisation increases its maturity by complying with additional KPAs and in the order fixed by the model. The initial level in the staged-view model is Level 1, which indicates that the organisation does not fully comply with the Level 2 KPAs.

At the lowest maturity level of the model, the software development process is implemented in an ad-hoc manner where different approaches are used for different software products. In many cases, budget and time estimates are not met. The quality of the products is unpredictable as it is based on an individual's effort, knowledge and experience. The organisation has to introduce basic management activities to rise to the second level of the model, the "Repeatable" level. At this stage, more uniformity is brought into the software development process in such a way that the organisation is capable of producing products in similar ways with similar quality. Quality assurance and configuration management are important for the transition from Stage 1 to Stage 2 in SW-CMM.

As the organisation rises to the third level, the "Defined" level, it should become capable of describing and managing its software development processes according to standards. As the organisation becomes familiar with the main parts of its standardised development processes, it can then control them. In order for the organisation to achieve the "Managed" level, it needs to quantify data regarding the software development processes. This data should be analysed and used in establishing estimates for planning and changes.

The highest maturity level of SW-CMM is the "Optimising" level. At this level, the organisation continuously improves its processes. This includes seeking and eliminating root causes of product and process defects, as well as incorporating new technology as appropriate. The organisation should be capable of developing processes as dictated by the characteristics of the product to be developed.

### Continuous-view architecture

In the SE-CMM, a continuous model, an organisation first achieves Level 1 in a process area by performing all of the process area's base practices in any manner. Not performing the base practices will earn the organisation a Level 0 – the initial level in continuous-view models – in that process area. To improve the capability of a process area beyond Level 1, an organisation should comply with additional generic practices in the way it performs the base practices.

For each level of the SE-CMM, the generic practices are based on the same principles as the corresponding level in the SW-CMM. SE-CMM Level 2 generic practices, for example, address project management discipline, as applied to a process area. This means that an organisation can achieve Level 2 in "Verify and Validate System", for example, if its projects perform all the process area's base practices and allocate adequate resources to perform the validation and verification activities. The organisation needs also to perform all of Level 2 generic practices such as assigning responsibilities for preparing and performing the necessary tests and analyses, and documenting the project's test and other verification processes.

If these project processes are tailored from an organisational standard set of test processes, using organisationally approved tailoring guidelines, and if the organisation uses well-defined data and performs appropriate reviews including peer reviews where needed according to Level 3 generic practices, then the organisation achieves Level 3 in "Verify and Validate System", and so on for the rest of the levels regarding all of the 18 process areas covered by the model.

### Staging and equivalent staging for continuous-view models

"Staging" is a concept by which a continuous-view model can be made to behave like a staged-view model (see Bate 1995; SEI 1999). A capability level profile is a list of process areas and their corresponding capability levels. The profile may be an achievement profile when it represents the organisation's progress for each process area while climbing up the capability levels. Or the profile may be a target profile when it represents an objective of process improvement. A target staging concept defines an order in which process areas of a continuous model are to be addressed and their associated requirements (generic practices) on capability levels to be achieved. Achieving those capability level ratings in that subset of process areas and associated generic practices earns the organisation an overall maturity level rating. Equivalent staging is a target staging that is defined so that the results of the target staging can be equivalent to the maturity levels of the staged representation. Such staging permits benchmarking of progress between organisations, enterprises, and projects, regardless of the CMM representation view used (SEI 1999).

### 8.7.4 People-CMM

In addition to the CMM, SEI also developed People-CMM (also known as P-CMM) (seeSEI 2001). The aim of this framework is to develop and manage the knowledge, experience and motivation of employees in the organisation. Just as in CMM, the maturity of the organisation, with regard to human resource management, is charted and improvement priorities are indicated.

The P-CMM, is an adaptation of SW-CMM concepts focused on developing the organisation's human capabilities, especially in software and information systems development. The motivation for the P-CMM is to radically improve the ability of software organisations to attract, develop, motivate, organise and retain the skills needed to steadily improve software development capability (SEI 2001).

The strategic objectives pursued in the P-CMM are to:

- improve the capability of software organisations by increasing the capability of their staff;
- ensure that software development capability is an attribute of the organisation rather than of a few individuals;
- align the motivation of the staff with that of the organisation;
- retain assets (i.e. people with extensive skills and capabilities) within the organisation.

Figure 8.5 depicts the key process areas within each maturity level and lists activities to be performed to progress between levels. As in SW-CMM, People-CMM consists of five maturity levels.

- Level 1 is the "Initial" level.
- Level 2 focuses on introducing basic discipline into workforce activities providing key process areas within work environment, communication, staffing, performance management, training and compensation.
- Level 3 is the focus on the identification of organisation's competencies and using its people's management activities to enhance them. The key process areas at this level are knowledge and skills analysis, workforce planning, competency development, career development, competency-based practices and participatory culture.
- Level 4 introduces a quantitative management view on improving people management capabilities and in establishing teams based on competence. Key process areas at this level are: mentoring, team building, team-based practices, organisational competency management and organisational performance alignment.
- Level 5 covers issues that address continuous improvement methods for competence development at an organisational and an individual level. Key process areas at this level are personal competency development, coaching, and continuous workforce innovation.

Table 8.2 lists the key process areas assigned to process categories. There are four of these categories: developing capabilities, building teams and culture, motivating and managing performance, and shaping the workforce. As shown in Table 8.2, the P-CMM includes the following activities in the defined KPAs for this model:

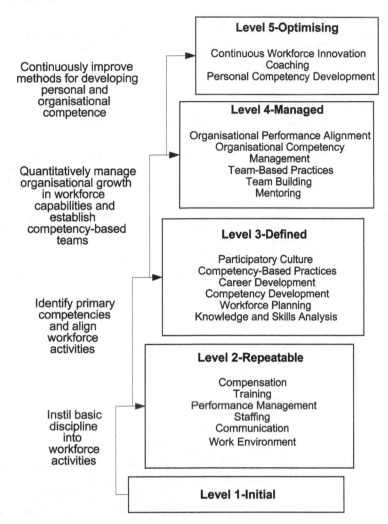

*Figure 8.5* P-CMM key process areas (SEI 2001).

- staffing (includes recruiting, selection and planning)
- managing performance
- training
- compensation
- work environment
- career development
- organisational and individual competence
- mentoring and coaching
- team and culture development

*Table 8.2* P-CMM key process areas assigned to process categories

| Category | Developing capabilities | Building teams and culture | Motivating and managing performance | Shaping the workforce |
|---|---|---|---|---|
| *Maturity levels* | | | | |
| **5. Optimising** | Continuous capability improvement | Optimise integration of processes | Organisational performance alignment | Continuous workforce innovation |
| **4. Managed** | Mentoring Competency-based assets | Empowered Workgroups Competency integration | Quantitative performance | Capability management |
| **3. Defined** | Competency development Competency analysis | Workgroup development Participatory culture | Competency-based practices Career development | Workforce planning |
| **2. Repeatable** | Training and development Compensation | Coordination and communication | Compensation Performance Management Work Environment | Staffing |
| **1. Initial** | | | | |

Source: SEI 2001

### 8.7.5 *Personal Software Process (PSP)*

The original drive for developing the PSP came from reservations raised regarding CMM. Many viewed CMM as being designed for large organisations, and did not see how it would be applied to individual work or to small project teams. As a result, PSP was developed to adapt 12 of the 18 CMM key process areas (Version 1.1) to improve the work of individual software developers. The purpose of PSP is to make individual software developers aware of their own software development process in order to improve the quality of that process, its predictability and the quality of the result. PSP is introduced on the basis of a training course which offers the opportunity to learn and apply PSP (Humphrey 1995).

PSP was based on the notion that improved personal process discipline can help to increase the effectiveness of individual software engineers. It was also felt that as the individual software developer's performance improves, the performance of the teams he/she is part of would also be more likely to improve. PSP is a self-improvement process designed to help individual engineers to control, manage, and improve the way they work. It is a structured framework of forms, guidelines and procedures for developing software. If properly applied, PSP provides the individual software engineer with historical data that helps in scheduling tasks and meeting them, which makes the software engineering work predictable and more efficient.

### Relationship between PSP and CMM

Organisations are generally interested in the personal process because they see it as an efficient way to bring process improvement benefits to small projects and organisations. Other groups see it as a way to avoid much of the cost and the organisational disruption of a major improvement effort. While PSP can greatly facilitate process improvement in both large and small organisations or even at an individual level, it should not be viewed as a replacement for an organisational process improvement effort. The two, in fact, are complementary. However, organisations near the CMM Maturity Level 2 – the "Repeatable" level – or above, are likely to be most successful in introducing the PSP. In contrast with the CMM, the PSP does not provide for an assessment technique to identify the PSP maturity level.

### 8.7.6 Team Software Process (TSP)

This model is introduced by SEI as an intermediate level, after CMM for the entire software development organisation and PSP for the individual software engineer, in which their processes can be mapped and improved (Humphrey et al. 1999). When organisations started process improvement from the initial level of any CMM model, the PSP leads software developers to how to address their tasks in a professional way. Once software developing teams are well acquainted with PSP, they need to start focusing on applying Team Software Process to their projects. TSP guides developing teams in launching, planning and managing their projects. Perhaps most importantly TSP shows managers how to provide leadership in guiding and training/coaching their teams to produce successful projects. TSP has five objectives:

1   To establish self-directed teams that plan and keep track of their work progress, establish goals, and control their plans and processes
2   To guide managers how to train/coach and motivate their teams and how to perform at their best
3   To speed software process improvement
4   To provide improvement guidance to the more mature organisations
5   To facilitate an educational system of industrial-grade team skills

TSP guides teams through the four typical phases of a software project. These projects may start or end on any phase, or they can run from beginning to end, as illustrated in Figure 8.6.

Before each phase, the team goes through a complete launch or relaunch, where they plan and organise their work. Humphrey et al. (1999) state that if

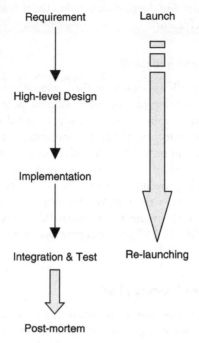

*Figure 8.6*  TSP structure (based on Humphrey et al. 1999).

team members are well acquainted with PSP it is adequate to have a four-day launch workshop which offers guidance to complete a project phase. In this workshop, TSP teams produce:

- written team goals
- team roles' definition
- process development plan
- team quality plan
- project support plan
- overall development plan and schedule
- detailed next-phase plans for team members
- project risk assessment
- project status report

For each of the subsequent phases a two-day relaunch workshop is needed to provide the team with guidance. In the final launch step the project team conducts a review of the plans and risks and conducts weekly meetings. The team also produces periodical status reports for project stakeholders.

The CMM, PSP, and TSP provide an integrated three-dimensional framework for process improvement (see Humphrey 1998a, 1998b, 1998c). CMM has

18 key process areas, and the PSP and TSP guide engineers in addressing almost all of them. These methods not only help engineers to be more effective, but also provide the in-depth understanding needed to accelerate organisational process improvement.

### 8.7.7 Bootstrap

Bootstrap is a normative process-based assessment approach for the improvement of the software development process of organisations (see Kuvaja et al. 1994). It was developed by an ESPRIT project sponsored by the European Commission from 1990 to 1993. The aim of the project was to develop a useful method for software quality improvement, to be European-cultured for the European market rather than the US-cultured CMM, ISO and others. One of the major differences between CMM and Bootstrap is the emphasis that Bootstrap places on the strength and weakness profile of an organisation in contrast to the maturity levels of CMM. A software-developing organisation does not need to master all processes to achieve a higher level. All processes are analysed to see if the organisation implements them adequately.

Apart from the primary development processes such as requirements of software development (analysis, design, implementation and testing), Bootstrap also investigates supporting processes (see Figure 8.7). Examples of these are project management, quality management, configuration management and in particular, the generic management practices such as the relationship with customers and users, human resource management and process improvement.

A major difference between CMM and Bootstrap is that CMM requires all processes at a given level and below to be implemented to achieve that level, while a Bootstrap level is achieved based on the average of multiple processes in an organisation. For example, if an organisation implements all but one of Level 2 processes and implements some of Level 3 processes, a Bootstrap assessment may measure Level 2, while CMM assessment will be Level 1.

### 8.7.8 Trillium

The Trillium model is used by Bell Canada to assess the product development and support capability of prospective and existing suppliers of telecommunications or IT-based products (Trillium 1996). It provides key industry practices which can be used to improve an existing process. The model is based on SEI's Capability Maturity Model (CMM) Version 1.1 and covers ISO 9000 series, among other standards.

The goal of the model is to provide a means to initiate and guide a continuous improvement programme. The model is used in a variety of ways: to benchmark an organisation's product development and support process capability against best practices in the industry. Also it can be used for

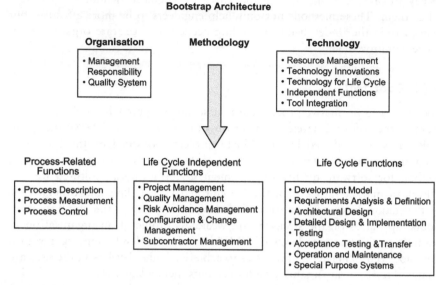

*Figure 8.7* Bootstrap architecture (based on Kuvaja et al. 1994).

self-assessment to help identify opportunities for improvement within a product development organisation, and in pre-contractual negotiations, to assist in selecting a supplier.

There are five maturity levels and eight capability areas within the Trillium model. Each capability area contains practices at multiple Trillium levels. For example, Management spans over Levels 2 to 4, while Quality System spans over Levels 2 to 5. To achieve a Trillium level, an organisation must satisfy a minimum of 90 per cent of the criteria in each of the eight capability areas at that level. Levels 3, 4 and 5 require the achievement of all lower Trillium levels (i.e. levels cannot be skipped).

### 8.7.9 SPICE and TickIT

An adaption of ISO 9000 standards specifically into the software industry resulted in the SPICE (Software Process Improvement and Capability Determination) (ISO 15504) (see El Emam et al. 1998) and TickIT models (see Hall 1997). SPICE is a worldwide collaborative effort to support the development of a new international standard for software process assessment. Using this model it is possible to report assessment results on a scale of capability from 0 to 5, with each level reflecting specific and incrementally significant issues in managing and improving software development and acquisition. A comprehensive series of user trials was incorporated into the SPICE plans.

Like SPICE, TickIT (Hall 1997) is a certification scheme but developed for the European market. It was developed by the British Department of Trade and Industry (DTI) and the British Computer Society (BCS) to apply ISO9001. The aim of developing this scheme was to create a detailed method for organisation, procedures and rules for a Software Sector Certification Scheme (SSCS) which would cover the assessment and certification of an organisation's software quality management scheme.

# Chapter 9

# Organisational readiness
# Bridging the gap

## 9.1 Gap analysis (performance analysis)

One of the most commonly used techniques for business improvements is gap analysis. This technique is based on two performance levels – current and target. The current performance level identifies the current status of the business function (or indeed any other organisational matter) in terms of the pre-identified evaluation criteria. The target performance level identifies the desirable status for this particular business function and is measured by the same evaluation criteria. The difference in the value of the evaluation criteria between the current and target performance levels is known as the "opportunity gap". This gap can provide an organisation with a focus and guidance on how best to progress from the current to target performance. The scale and importance of this gap enables managers to assess the subsequent likelihood of this affecting the key business strategy deliverables, thereby enabling corrective action to be taken as appropriate.

In addition, gap analysis is a technique that can be used for measuring performance achievement against set targets. It is particularly useful for measuring outputs of an activity over a period of time.

For example, Figure 9.1 highlights three main areas for performance analysis: Area "X", Area "Y" and Area "Z". The performance target achievement levels

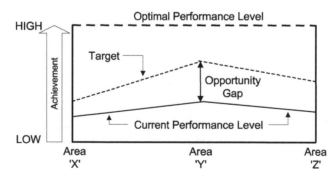

*Figure 9.1* Performance analysis chart.

for each of these areas are represented by the dashed line. From this representation, it can be seen that the greatest area for improvement rests with Area "Y". In this context, if this activity was classed as critical to the delivery of the business strategy's objectives, then resources should be prioritised accordingly to this area in order to address performance deficiencies and close the identified performance gap.

Also this technique can be utilised to address cultural problems in organisations brought about by employees' deep-rooted values and beliefs. The term "culture gap" is the difference between the values and perceptions held by one person, group, or subset and the perceptions held by others (individuals, groups, or subsets). For example consider the views of management and operatives: each have different perceptions of culture and of each other, but neither is able to identify why this gap exists, how wide it is, or how this gap can be closed. If such a gap is not effectively addressed, these fundamental differences can often instil mistrust and create communication problems, particularly where no common agreement has been reached concerning social rules and protocol.

Gap analysis is a valuable tool used for controlling resource-driven activities over time (to predict trends), the mechanics of which can often be applied to a variety of different process areas. Whilst these areas usually embrace production-related activities, the analysis can also be extended to include skills development (Van Daal et al. 1998), intellectual capital (Joia 2000), gaps in IS and IT (Ward and Griffiths 1997), gaps in culture (Ward and Griffiths 1997) and even human resource management (HRM) issues. Furthermore, from a strategic perspective, it can also be used to identify gaps in organisations' capabilities in relation to the opportunities and threats associated with the external environment (Harrison and Pelletier 2000).

The accuracy of the gap analysis technique is highly dependent on how the business function under consideration is measured and what criteria are used, not only to measure its current performance but also to measure progress of improvement on the path of achieving the target level. The following section briefly explains three of the available techniques that can assist organisations to identify the opportunity gap.

## 9.2 Measuring the opportunity gap

### 9.2.1 Benchmarking

Benchmarking is a process that attempts to measure the performance of one company (or activity) against the performance of another. It is thought to have originated in the 1970s with the Xerox Corporation in the USA, where it helped the company improve competitiveness and stay in business.

The main rationale of benchmarking, however, is to establish and measure "best practice" in processes deemed critical to the delivery of BS objectives. It can be used as a catalyst for change (see Pickrell and Garnett 1996), and if used appropriately, can help organisations manage the transition of change more effectively. Therefore, it is a dynamic and evolving process, the results from which can enable organisations to continuously develop and evolve, and to generally perform better in the marketplace. In addition, it also allows performance gaps to be identified (see Karim et al. 1997; Mohamed and Tilley 1997). However, the actual procedure of benchmarking can only be undertaken if the processes to be benchmarked are fully understood (and are capable of being measured against).

Benchmarking is normally undertaken with competitors in a similar environment and market sector ("best in class" or "world-leaders" if possible), therefore enabling cognate comparisons to be made which will facilitate continuous improvement (Davenport 1993). In addition, there is a school of thought that suggests that benchmarking should also be applied across other industry sectors irrespective of discipline (Clark et al. 1999; Atkin 2000), which could offer complementary insight (and corresponding business benefits), especially where generic processes exist.

Users of benchmarking must however be aware of its limitations, as the factors that influence success are not always readily available. Furthermore, the choice of company to benchmark against can also provide a further set of difficulties, as narrow differences can often lead to misinterpretation of findings. This can also be adversely influenced by ethical and legal issues (Spendolini 1992), or through insufficient knowledge of the exact processes being benchmarked against.

### 9.2.2 Balanced scorecard system

The balanced scorecard (BSC) is a conceptual framework developed by Kaplan and Norton (1992) to assess the impact of decisions on the business. It is used to translate the corporate mission and strategic objectives into a set of performance indicators. These performance indicators are divided into four perspectives.

- Financial – targets, profit turnover, etc.
- Customer – quality, service, stakeholder needs, etc.
- Internal business processes – processes, performance, objectives, etc.
- Learning and growth – skills, capabilities, change issues, alignment, etc.

The BSC methodology enables users to set targets and measure progress for each of these four perspectives, in an attempt to overview and determine the company's position. This process identifies areas for improvement, and places "value" on results. All targets have milestones, the structure of which establishes the strategic direction. The BSC can also be used to accommodate generic performance measures matched to corporate goals. Later refinements to this framework focus on using the BSC as a strategic management system (Kaplan and Norton 1996). These developments invite managers to concentrate on improving or re-engineering those processes deemed critical to organisational strategic success. It also enables the BS to be aligned to predetermined "action" criteria.

### 9.2.3 SWOT analysis

The SWOT (strengths, weaknesses, opportunities and threats) analysis is an analytical tool used to form (and shape) the business strategy. The procedure is also known as "situation analysis". Its main purpose is to identify strategic pointers, to focus resources and capabilities effectively (Robson 1997). It normally considers the following issues:

- Resources – financial health of the company, etc.
- Products offered – appropriateness, etc.
- People – skills, competence, leadership, motivation, etc.
- Assets and liabilities
- Organisational structure – corporate strategy, level of culture, etc.
- Market influences – present and future
- Competitors – effect on organisation

These issues are condensed into the SWOT analysis, allowing the organisation to use its strengths to exploit opportunities, address weaknesses and defend against threats (Ward and Griffiths 1997). Strengths include issues that the organisation is good at, such as core capabilities, or a distinct product. Weaknesses, however, include items that the organisation needs to address, covering resources issues, skill levels, inability to respond to change, etc. Opportunities are areas that the organisation could exploit, which often includes new markets, mergers, etc. Whereas threats are the immediate dangers facing the organisation, which tend to include any immediate and pressing issues, such as competitors (but can also embrace technical, economic, physical, political and social factors).

The SWOT analysis can provide organisations with a structured and systematic method for clarifying purpose (to achieve business focus and direction). Furthermore, additional benefits from this approach can be derived by co-joining this technique with other business performance measures (Lee and Sai On Ko 2000). However, managers should be aware that, whilst core capabilities and resources should be coordinated to support and underpin the business strategy, this must be tempered by balancing opportunity with competence (K.R. Andrews 1987).

## 9.3 IS life cycle and organisational readiness

### 9.3.1 Life cycle of IS

In Chapter 7 it was explained that IS can potentially bring about improvements and strategic advantage to organisations if IS strategies are aligned with business strategies. A number of tools already exist to assist organisations in developing IS strategies in support of the business strategies. (This is the first phase of initiating information systems in organisations. See Figure 9.2). For example, organisations can adopt strategic information system planning (SISP) to identify and prioritise IS that can best support and meet their business objectives. This task is either carried out by the organisations' IT departments, if there is an in-house capability to do so, or by specialised consultant organisations. This phase leads to the identification of business-critical information systems which are necessary to improve business performance or create a competitive advantage for the organisation. Once this phase is completed and the recommended information systems are approved and budgeted for, the development and implementation phase normally commences under the control and management of the organisation's IT department. At this phase the focus of the organisation shifts to ensure that the new system is successfully developed (or customised, if it is a third-party product), implemented within time and budget, and that it meets its predefined objectives (see Figure 9.2).

The development and implementation phase is a "product oriented" phase where the main focus of the development team is on issues critical to the development and delivery of the product such as project management, development process, aligning business processes with the system's functions, user satisfaction and resource utilisation. The implementation process of the new system can either follow a "big bang" approach or a gradual approach to minimise risk. Once it is implemented, the system is then assessed to examine its value and contribution to the organisation by adopting approaches such as those explained in Chapter 3.

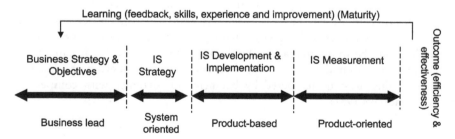

*Figure 9.2* IS lifecycle phases.

## 9.3.2 IS success and organisational readiness

The above approach to the life cycle of IS has been followed since the 1970s. Clearly the effectiveness of this approach is challenged due to the high percentage of IS failure, i.e. those projects that failed to achieve their intended business objectives (as explained in Chapter 3). As previously mentioned, the main reasons for systems' failure are related to organisational soft issues such as business processes, management structure, change management, people and culture. This implies that organisations were not in a state of readiness to allow them to absorb the new systems into their current work practices. Typically, new systems are "alienated" from organisations' work environment either during the development and implementation phase and therefore disregarded before completion, or after the completion of this phase and therefore resulting in unused systems. To address this problem effectively, an appropriate organisational culture and learning environment must exist prior to the commencement of the system's development and implementation phase, in which information systems are considered as essential to the business and an integral part of the organisation's resources infrastructure – particularly by its top management – and are supported by a flexible work environment which is ready to accept change (see Chapter 5). This environment is necessary for the successful implementation of the newly proposed information systems and also has a significant role to play in initiating innovative information systems in support of the business objectives, i.e. influencing the outcome of the first phases of the IS life cycle (see Chapters 5 and 8).

The resource-based models further explain the above facts and show that if an organisation is to develop an IS capability it has to have competencies to pull the right resources together in the most efficient way within the organisation. This involves the creation of a responsive work environment with a high level of experience in process management, skilled and empowered employees, and appropriate management structure and schemes. Hence, successful implementation of IS requires an organisation to carefully consider how best a new system can be fully integrated into its operations and whether the organisation has the capacity to do so. In this context it is important that organisations identify their level of readiness to integrate new IS into their current operations prior to committing large investments into the development and implementation of these systems.

There is a general agreement that the two main factors influencing the successful implementation of a new IS in organisations are business processes and people. These factors need to be carefully assessed by managers prior to commencing the development and implementation phase. Business processes need to be mapped (aligned) with the proposed system's "built-in" functions as shown in Figure 9.3. This mapping activity could:

1   disturb the current performance of the organisation;

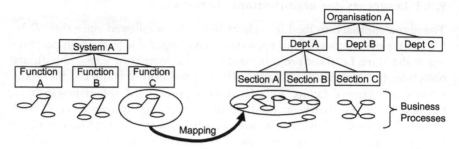

*Figure 9.3* Mapping systems' functions onto business processes.

2    require large-scale change of existing processes and thus be expensive and risky to carry out;

3    face high resistance from employees.

Moreover, the level of impact of this mapping activity on the success of the new IS is highly affected by the current work environment and IT infrastructure of the organisation, as explained in Chapter 2, Section 2.3. The work environment is influenced by factors such as management structure, current practices, motivation and empowerment of employees, and roles and responsibilities. The higher the ability of the organisation to adapt to the new work practices, the more likely that the organisation can successfully integrate the new IS into its operations.

On the other hand, the current level and type of IT infrastructure used in an organisation can influence the success of embracing the new IS into the organisation. This factor does not only reflect the current technical capability of employees but it can also be used as an indicator of the level of IS/IT awareness and expertise of the employees. The longer and more strategic the use of IS/IT in the organisation, the more aware and competent the work force is and thus the easier it is to embrace the new system (see Chapter 2, Section 2.4.2). In this case, employees would have developed the necessary competencies out of using the current IT infrastructure, which would enable them to adapt to the new system with minimum resistance. Such skills and competencies can be the main catalysts for the establishment of an organisational learning environment (see Chapter 5).

It has to be pointed out that the risk posed by processes and people to the successful implementation of a new IS depends on the size and type of the proposed system. Systems that are fundamental for sustaining competitive advantage are more likely to cause greater change to organisations compared to those of an operational nature. For example, adapting a cross-functional system to integrate core business functions – e.g. integrating design and construction functions or integrating estimating and planning functions – may not only demand a significant deviation from the current practices, but may also demand new procurement methods that have to address the new functionalities introduced by the new system taking into consideration the new roles and responsibilities of each party. On the other hand, implementing an operational system – e.g. a CAD system – to improve the efficiency of a particular design task may not require a critical change. In the latter case, the main demand will be to acquire the necessary skills through technical training to successfully run the system.

In summary, the capability of an organisation to integrate a new system into its current operations depends on its level of readiness to:

1   adapt to the new business processes;
2   embrace the new business process (awareness, competencies and skills of employees);
3   adopt the changes into the organisation's work environment (a work environment which eases and encourages the expected changes to occur);
4   accommodate the new technology within the existing IT infrastructure and management.

The following section further discusses the relationship between the above factors and proposes a "measurement of readiness gap" phase to the IS life cycle, i.e. between the business and IS strategy phases and the development and implementation phase.

### 9.3.3 Measuring organisational readiness

IS measurement and evaluation are mainly focused on measuring the impact of an IS "as a product" on the level of user satisfaction after the systems' development and implementation phase is completed. More recently, however, efforts have started to shift away from this approach by focusing on the recipient organisation as being the main beneficiary. Such efforts concentrate on organisational soft issues, i.e. business processes, people, work environment and IT infrastructure. However, clear and easy approaches are yet to be developed with the aim of measuring these four elements and incorporating them into formal guidelines which managers can use to assess the organisational readiness gap prior to the commencement of new IS investment, i.e. the development and implementation phase.

The indication is that adopting such approaches can seriously affect the IS investment decisions and hence the start of the development and implementation phase of the IS life cycle. This approach can be incorporated into the IS life cycle through the addition of a new phase "measuring readiness phase" (see Figure 9.4). This is an organisation (business) oriented phase which needs to be carried out after the completion of the IS/IT strategy phase or after a particular IS system is selected for implementation, and before the commencement of the development and implementation phase. This new phase will provide

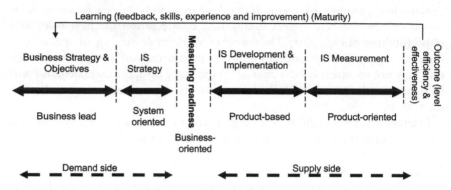

*Figure 9.4* Introducing "measuring readiness" phase.

organisations with an early warning of the risk level that they might face during the implementation of a new IS and therefore will act as a risk measurement buffer. It will also allow organisations to identify and estimate the required level of organisational change in order to successfully implement the selected IS. If this approach is successfully followed it will minimise the required management of change and people's resistance to the new system, thereby creating an essential work environment for the success of integrating the new systems into the organisation's work practices.

To adopt the above concept, Figure 9.5 shows the sequence of events that needs be followed prior to the IS development and implementation phase. At the completion of the strategic planning and the identification of the IS applications that are required to support the business strategy, the capability assessment of the organisation to successfully implement the proposed IS should be carried out. This involves the assessment of four key elements (process, people, work environment and IT infrastructure), which determines the level of organisational change required in these four elements. This assessment task must be assisted by a management tool to identify the readiness gap in each of the four key elements, i.e. to identify the difference between the current and target levels of each element. When the readiness gap is identified, a key decision has to be taken by the organisation to either proceed to the development and implementation phase (if the readiness gap is small) or to plan for organisational improvements with the aim of reducing the readiness gap (if the readiness gap is too big).

### 9.3.4 Maturity models and measuring the readiness gap

The concept of maturity models, as explained in Chapter 8, can be adopted to provide the basis for measuring organisations' IS readiness in the above four key elements. This can be achieved through the establishment of a number of specific measurable attributes, for each of the four key elements, where the

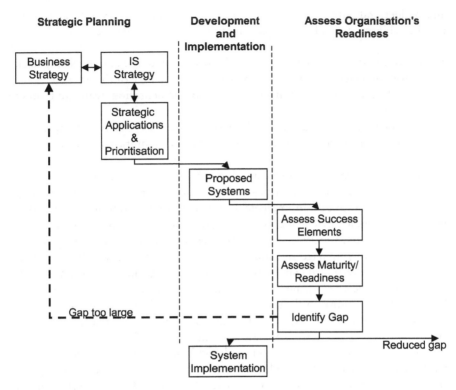

*Figure 9.5* Assessing organisational readiness.

maturity level can be identified. The current level of a key element defines its existing status in the organisation while its target level defines the "to be" status of that element prior to the implementation of the new IS. Each attribute should have the capability to be described, at any maturity level, by certain criteria. The higher the level of maturity of the element, the better and more flexible its attributes are towards the successful implementation of the new system. Progressing from one maturity level to another should be guided by a Readiness Maturity Model which will provide organisations with guidelines on the required improvement for each of the four elements.

## 9.4 IS assurance

Another important issue that needs to be carefully considered by organisations prior to the commencement of the IS development and implementation phase is the quality assurance for this phase. Upon the selection of a proposed IS and the approval of the required budget for its implementation, top management needs to be assured and kept informed during the entire development and implementation phase that this phase will achieve its objectives within the estimated time and budget and with minimum risk. Quality assurance or "IS Assurance" can best be achieved through the establishment of an independent team to:

- continuously assess the progress of the project with the aim of feeding back to top management on the progress and performance of the project;
- provide an early risk warning to the implementation team and other stakeholders;
- provide independent advice on the actions recommended by the implementation team to keep the development and implementation phase on track.

The development and implementation of large and complex systems, in support of core business functions, is normally contracted to a third party (software vendor and/or an implementation consultant). Internally, an implementation team, led by members of the organisation's IT Department, is established to guide and supervise the third parties' activities. The capability of this team to effectively manage and control such a critical task is highly dependent not only on the technical capability of the IT department but also on its capability to handle the organisational change as required by the new system. Lacking the capability to do this effectively, in most cases, will create a gap between the consultants and the organisation. This will eventually lead to the consultants' role shifting from an "advisory and training" one to a "leading and driving" role. This can bring about a difficult situation for the organisation particularly as the recommended solutions (mainly business processes) may not suit its own culture and thus the implemented system risks being turned into a failure. (See Chapter 2, Section 2.4 for further details.)

However, the risk of IS investment can be effectively managed if an independent IS assurance team is established to fill the gap between top management, the IT department and the external consultant/vendor. The team should be carefully selected with roles and responsibilities clearly communicated to all the project's stakeholders. This task needs to be carefully considered as it can have

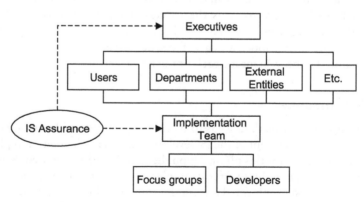

*Figure 9.6* IS assurance team and project stakeholders.

a negative affect on the performance of the consultants and the implementation team if these parties are not fully aware and appreciate the role of the IS assurance team. Simply, it can be seen to undermine the capability and authority of the IT department as well as tending to hinder the progress of the consultants. It is therefore highly important that when such a team is created – in itself a big investment for the organisation to undertake – the roles and responsibilities are clearly set and that a clear "mindset" is established between all the project's stakeholders leading to building trust and confidence in the IS assurance team. Figure 9.6 shows an example of how the IS assurance team can be embedded into the project's implementation structure

## 9.5 IT training strategy: the GAPP-IT model

In Chapter 7 the importance of human capacity building to organisations' continuous improvements and competitive advantage was clearly demonstrated. This section explains a model which was developed to facilitate IS/IT capability building in organisations through the adoption of "performance gap" principles. The performance gap analysis (see Section 6.7.6) highlights three skills categories that need to be addressed, namely operational, managerial and executive. Furthermore, the training needs for these categories must be aligned with the organisation's business strategy to ensure that the organisation has the human capability to successfully achieve its business objectives. The GAPP-IT model (Generic assessment process protocol model for IT training) was developed to determine and "map" the key generic processes associated with IT, IT training and the impact on the BS (Goulding and Alshawi 2002).

### 9.5.1 Process protocol principles

The GAPP-IT model adopts the principles of the Process Protocol (PP) modelling (Kagioglou et al. 1998). The latter is capable of presenting all diverse parties interested in a process in a flexible, clear and standardised framework of generic activities. This framework encourages professionals to understand and appreciate the process more easily, affording improvements in communication and co-ordination, the control and management of resources and the adoption of a "shared vision". The generic nature of PP can be applied and adapted to suit many disparate and diverse project environments. Cooper et al. (1998) demonstrate the application of PP to the generic design and construction processes.

The key attributes of this framework are:

Activity zones – a structured set of sub-processes designed to support the solution.

Deliverables – outputs from project and process information, used to create the *Phase Review Report*.

Phase review and stage gate processes – generic processes within the stages, separated by decision gates (*Phase Review Meetings*) needed to fix and approve the information prior to progression.

Gates and phase reviews – project and process review points used to examine progression and dependant upon predefined criteria. Gates are either *hard* (prevent progression) or *soft* (accept conditions and allow concurrency).

Legacy archive – a mechanism for storing, recording and retrieving project and process information.

*Phase Review Report* – documentation of deliverables presented at the phase review gates: this information is then subsequently stored in the legacy archive.

### 9.5.2  The development of the GAPP-IT model

The GAPP-IT model, developed in collaboration with two major UK construction industrial partners (supported by domain experts, process stakeholders and specialist advisers), aims to analyse the key sequential stages (processes) required for evaluating the impact of IT training on the business strategy of an organisation.

The first step in the development of this model had to identify and define the target audience where training needs were divided into three discreet categories – operational, managerial and executive. These categories were chosen to differentiate between the different types of skills sets needed. Subsequent stages considered the outline stages (processes) and sequence of events required for instigating and deploying IT training within a construction environment. This included all contextual issues associated with training, from inception through to evaluation and final reflection. Consequently, this process needed to appreciate (and embrace) all human resource enablers and business process issues into one holistic PP framework (see Goulding and Alshawi 2002).

Seven key phases are offered for discussion, encompassing the initial preliminary stage (conception of the need for IT training), through to the final evaluation and feedback phase, as presented in Figure 9.7. Each of these phases can be broken down to show lower level information and processes.

### 9.5.3  The GAPP-IT model

The GAPP-IT model (see Figure 9.7) is divided into three horizontal levels covering the operational, managerial and executive requirements. It is commenced in Phase ZERO, and exited in Phase SIX, after sequential progression through each phase has been achieved (if required). The gaps between each phase signify the presence of a physical boundary or barrier (termed "stage gate"). Stage gates can be either hard or soft. Their classification is determined by the requirements and conditions set by the preceding/succeeding phases. Upon completion, users return to Phase ZERO through the feedback loop (via the process wheel). A brief summary of each of these phases is given below.

Phase ZERO  The aim of this initial opening phase is to establish the need for IT training from an outline perspective. It is used to evaluate the existing and future business needs, contemplating the current and future IS and IT strategies (and the potential changes they may have on resource requirements). The BS is agreed and confirmed in this phase, the structure and drivers of which help to determine the overall IS/IT demand. Users subsequently pass through a soft stage gate into Phase ONE – to formally identify and clarify these needs.

Phase ONE  This phase uses the information created in Phase ZERO to establish the processes involved in forming a generic IT training strategy. It is used to identify the BS deliverables, and to determine the precise scope and nature of the IS/IT demand to deliver these needs. A skills audit (see Chapter 6, Section 6.7.6) is used as part of this process to verify the existing skillS levels and to determine the subsequent type (and level) of training required. At this stage, it is possible to assess whether IT training is needed. If training is not required, users exit this model through a decision icon into the stage gate, where a *Phase Review Report* is completed (which is subsequently stored in the legacy archive). However, if training is required, an outline generic IT training strategy is formed at this point. Users then pass through a soft stage gate (as no financial resources have been committed yet) into Phase TWO.

Phase TWO  The exact corporate generic IT training needs are established in the form of a structured training plan (TP) in this phase. This requires financial resources to be determined for the delivery of all operational, managerial and executive generic IT training needs. Corporate funds are then prioritised to those areas perceived to have the greatest impact on the performance gap. Groups are subsequently informed of this generic provision – allowing them to make appropriate arrangements for their specific IT training requirements. However, as financial commitment is to be made during this phase, automatic progression is prevented due to the presence of a hard stage gate. In this context, a Phase Review board meeting is required to approve and sanction the anticipated resource expenditure and to agree the scope and content of the TP. A *Phase Review Report* is completed and users exit this phase into Phase THREE.

Phase THREE  This phase is used to establish the training and control mechanism to ensure critical needs, deliverables and deadlines are all met.

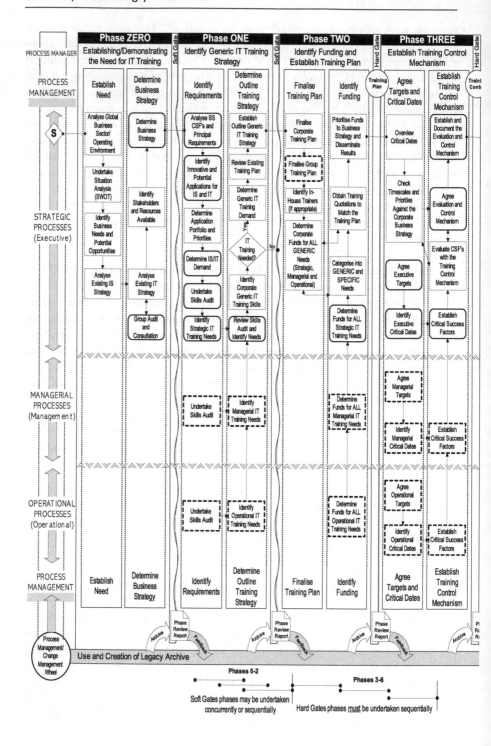

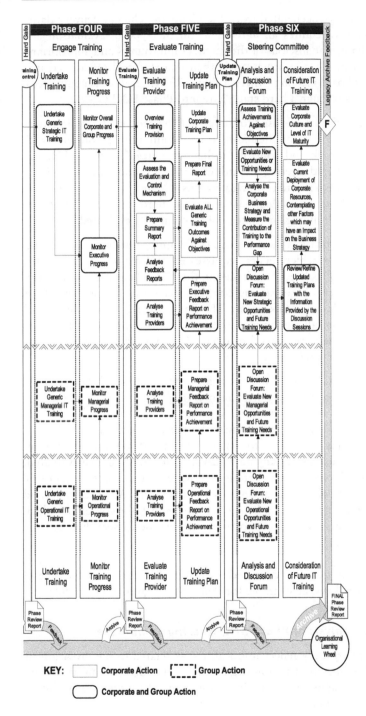

*Figure 9.7*  GAP-IT, the six phase model.

All IT training needs generated from the BS are matched to delivery dates. This process also requires an agreement on the type of evaluation and control mechanism (conscious of the impact on resources and appropriateness to the task). This information is then reported and sanctioned in a Phase Review board meeting (where a *Phase Review Report* is completed). Users then pass through a hard stage gate into Phase FOUR.

Phase FOUR  This phase is used to undertake the generic IT training identified. It is also used to monitor and control training in accordance with the training feedback and control mechanism agreed in Phase THREE. A Phase Review board meeting is conducted to assess and ratify all training achievements (and to record training progress to date), and a *Phase Review Report* is completed. Users then pass through a hard stage gate into Phase FIVE, where these training outcomes are subsequently evaluated.

Phase FIVE  This phase is used to evaluate training experiences and outcomes. All training achievements are assessed and measured against the original training objectives (established in the TP). The process also evaluates the effectiveness of the training control mechanism. The existing TP is subsequently updated at this point. The Phase Review board meeting is conducted to record all outcomes and achievements in the form of a *Phase Review Report*. Users then pass through the final hard stage gate into Phase SIX – the feedback stage.

Phase SIX  This is the final (and most important) phase of the GAPP-IT model. It uses a steering committee to overview and assess the whole process of IT training at the corporate and group levels. It is used therefore, to evaluate training achievements against the performance gap. An "open" discussion forum is undertaken to foster and stimulate discussion on training and development issues. Any new ideas or initiatives are then evaluated, contemplating the company's current deployment of resources, level of IT maturity, and prevailing level of organisational culture. All discussions and outcomes are documented in a *Phase Review Report*, the content of which also records process issues for improvement purposes. Users then exit the GAPP-IT model at this stage.

Each of these phases can be identified in further detail to show lower level process information flows, an example for Phase SIX is given in Figure 9.8.

### 9.5.4 Detailed analysis of Phase SIX: steering committee

Phase SIX aims to critically reflect upon all issues related to the provision of generic IT training in a construction organisation. It is divided into two main process streams: "Analysis and discussion forum" and "Consideration of future IT training". The details are shown in Figure 9.8.

Phase SIX is commenced in the "Assess training achievements against object-ives" process box. At this point, corporate and group executives focus on assessing the findings produced in the final report and updated corporate TP presented in Phase FIVE.

Particular attention is placed on examining the final training outcomes against the original objectives – to determine the achievements secured and how to address deficiencies (if evident). This information is then sent to the sub-process activity "Update skills register". Its details are used to update and maintain the corporate and group skills base. These findings are processed in the "Update personnel records" sub-process area, where all these files are updated. This procedure not only records achievements and continuing profes-sional development (CPD) information, but is also used to identify any future personal development areas for the next appraisal period. Upon completion, confirmation is sent back to the "Assess training achievements against object-ives" process box. This allows the next stage to be commenced: the "Evaluate new opportunities or training needs" process box.

The "Evaluate new opportunities or training needs" process box is used to identify further training areas or new initiatives that have arisen from the gen-eric IT training sessions, as the acquisition of skills and competence through the application of training can often lead to new business opportunities being identified. Typical areas often cited include innovation (through increased understanding of process), or new methods of working (with improved com-munication and level of integration). These opportunities should be developed and outlined in principle (at corporate and group level), and then subsequently forwarded to the "Analyse the corporate business strategy and measure the contribution of training to the performance gap" process box for further analysis.

The latter process box is the main area where IT training's contribution to the business performance gap is evaluated and assessed. At this juncture, all BS critical success factors (CSFs) identified in Phase ZERO and Phase ONE are systematically scrutinised in context with the deliverables achieved. This procedure is assisted by two sub-process activities: "Select timeframe and identify areas for assessment" and "Measure outputs".

The "Select timeframe and identify areas for assessment" sub-process activity is used to refine the areas allocated for assessment, and to define the timeframe used to measure achievements. Once this timeframe has been agreed, it is sub-sequently sent (along with the areas identified for assessment) to the "Measure outputs" sub-process box for analysis. This "Measure outputs" sub-process box is used for defining and measuring outputs. These outputs are generated from statistical evidence. Their results should be presented in the most appropriate way, e.g. graphs, tables, charts, etc. The output medium should be uniform to enable comparisons to be made, and clearly indicate performance achievement levels and variances from the expected norm. These findings are then directed to the "Analyse the corporate business strategy and measure the contribution of training to the performance gap" process area for final analysis and reflection. Key findings from this stage are subsequently processed and sent to the "Open discussion forum: evaluate new strategic opportunities and future training needs" process area for further discussion.

The "Open discussion forum: evaluate new strategic opportunities and future training needs" process box is used specifically for analysing all training findings (from a strategic perspective). This is undertaken using an "open forum" format, whereby all corporate and group stakeholders are able to positively contribute and discuss key issues and findings secured to date. This forum is not however undertaken until two other identical forums have been accomplished at the operational and managerial levels as indicated by the process boxes "Open discussion forum: evaluate new operational opportunities and future training needs", and "Open discussion forum: evaluate new managerial opportunities and future training needs" respectively. The information generated by these sessions can be fundamental to the progression of the strategic meeting, as these findings often shape (and influence) the final strategic decision-making process.

The operational and managerial open discussion sessions are used therefore to gain insight into "grass root" perceptions and beliefs (the equivalent of a conventional "suggestion box"). These sessions offer stakeholders an opportunity to contribute directly to the business, the consequence of which could lead to new opportunities being developed and exploited, or may highlight more efficient methods of working or improvement measures. These sessions are assisted by two sub-process activities to help facilitate and direct outcomes, specifically "Assess impact on company performance and document potential benefits" and "Prioritise".

The "Assess impact on company performance and document potential benefits" sub-process box is used to test ideas and concepts generated by the discussion sessions with seniors and peers. These findings are refined and subsequently prioritised in the "Prioritise" sub-process box, and are then redirected to the discussion group for final reflection. All findings generated from the "Open discussion forum: evaluate new operational opportunities and future training needs" process area are then sent to the "Open discussion forum: evaluate new managerial opportunities and future training needs" for evaluation. The combined results from these sessions are then refined and forwarded to the "Open discussion forum: evaluate new strategic opportunities and future training needs" process area for strategic analysis. When all data, ideas and concepts have been evaluated at the strategic level, the combined findings are then directed to the "Review/refine updated training plans with the information provided by the discussion sessions" process area for final reflection. The main task is to update and refine the group and corporate TPs. This is undertaken with two sub-process activities, specifically: "Assess areas most likely to improve company performance", and "Document findings".

The "Assess areas most likely to improve company performance" sub-process box is used specifically to identify "quick wins" or factors perceived to make an immediate impact on business performance. This process activity also investigates high potential areas deemed likely to have an impact on closing the performance gap. These findings are then sent to the "Document findings" sub-process box. The "Document findings" sub-process box is therefore used to document and prioritise all evidence. The confirmation details are then redirected to the "Review/refine updated training plans with the information provided by the discussion sessions" process area for ratification. The updated TPs are subsequently forwarded to the "Evaluate current deployment of corporate resources, contemplating other factors which may have an impact on the business strategy" process box for evaluation. This process area is thus used for assessing the impact and subsequent "value" derived from the generic IT training sessions from a corporate perspective.

From a resource perspective, executives often have the opportunity to direct corporate resources in a variety of different ways. These decisions are normally directed towards achieving maximum benefits (tangible or otherwise). For example, should resources be used to upgrade a computer network, purchase new software, or perhaps recruit more staff? Irrespective of outcomes, a formal strategy must exist that is capable of rationalising data into decisions that can be justified accordingly. Thus, part of the discussions within the "Evaluate current deployment of corporate resources, contemplating other factors which may have an impact on the business strategy" process box would debate these issues in depth. All interim findings from these sessions are then sent to the "Identify and prioritise all factors that contribute to the delivery of the BS CSFs" sub-process box for evaluation. In this sub-process arena, anticipated resource issues are matched to expected outcomes. For example, allocating "x" pounds to "y" area, will have "z" probability of achieving success. These scenarios should however be flexible enough to encompass subjective (or unconventional) outcomes. These findings are then redirected to the "Evaluate current deployment of corporate resources, contemplating other factors which may have an impact on the business strategy" process box for final reflection and subsequently forwarded to the final process box "Evaluate corporate culture and level of IT maturity" for closing analysis.

The "Evaluate corporate culture and level of IT maturity" process box is used to evaluate the company from a holistic perspective. This is achieved by assessing overall corporate and group performance, paying particular attention to factors (positive and negative) that influence outcomes. Part of this evaluation process should also endeavour to determine the change in corporate culture (if any) and overall employee willingness to accept and respond to change initiatives. The latter elemen,t in particular, is undertaken in the sub-process activity "Determine success of change management initiatives". Another fundamental part of this evaluation process is undertaken in the "Check IT and process maturity" process box. The level of IT maturity is assessed (from a CMM perspective), placing particular importance on the impact this technology has made on the business from a strategic and support perspective.

The final stage in the GAPP-IT model culminates in the "Evaluate corporate culture and level of IT maturity" process box. At this point, all results and findings are documented in the final *Phase Review Report*. Users then progress through a final hard stage gate (as the steering committee has been completed) and the model is subsequently terminated at this point. However, for the purposes of reflection and review, all *Phase Review Reports* are then directed through the Legacy archive conduit into the "Organisational learning wheel" and "Process management/change management wheel". These mechanisms are used to adjust and re-align the mechanics of the GAPP-IT model to match company-specific requirements. All process-related issues are then re-visited (by the process manager) to reflect best practice and improve the organisation's quality assurance systems and procedures.

### 9.5.5 Summary

The generic nature of this model is designed specifically to accommodate small to medium enterprises (SMEs) through to large construction organisations. The contribution of IT training to the "performance gap" can be determined using this model, as well as helping to assess the need for and impact of IT training initiatives on the BS. Moreover, investment decisions can be evaluated against the performance gap using targets that measure IT training's contribution to the delivery of BS key performance indicators. In this context, this model can also be used to improve process by capturing and sharing best practice to facilitate continuous process improvement. This philosophy is in line with the concepts of organisational learning (OL) as identified by Argyris and Schön (1978), Senge (1990), Huber et al. (1991) where the collective actions of individuals are structured to improve organisational knowledge and performance.

From an organisational performance perspective, construction companies can benefit from understanding how skills contribute to company performance (especially IT skills). However, it is important to note that the effectiveness of any IT training initiative can be influenced by many factors, not least the prevailing level of organisational culture and overall commitment to training. Furthermore, the use of CMM concepts (see Chapter 8) can be used to highlight progression stages, from "immature unrepeatable processes" through to "mature well-managed processes", the results from which can help organisations improve their appreciation of process maturity. Finally however, it is important to note that all findings and data generated from the GAPP-IT model should be routinely and systematically updated, as inaccurate (or out-of-date) information can adversely affect outcomes.

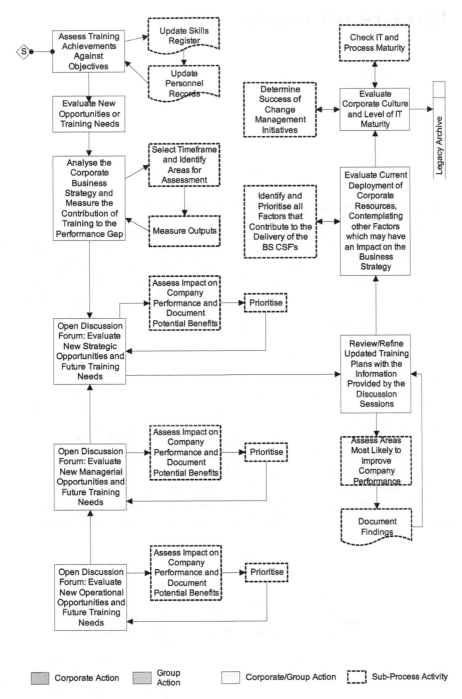

*Figure 9.8* Detailed phase attributes.

Chapter 10

# IS/IT readiness model

## 10.1 The Need for an IS/IT readiness model

Chapter 9 discusses the impact of the organisational factors on the successful implementation of IS/IT projects. It highlights the need for a management
tool to allow managers to assess their organisation's state of readiness prior
to IS/IT investment, i.e. prior to embarking on the development and implementation phase of the IS life cycle. Such a tool should have the capability
to identify and measure the current and the target status of organisational
readiness to successfully absorb new information systems into their work
practices.

Readiness models should have appropriate structures to embrace the four
key elements: process, people, IT infrastructure and work environment (see
Figure 10.1). Levels of readiness (both the current and target) should be clear
and indicative of the organisation's situation in terms of measurable attributes
and in maturity-like levels. Also, the model should be used at the proposed
new phase in the IS life cycle "Measuring readiness" (see Chapter 9, Section
9.3), i.e. when a specific IT/IS project has been selected by an organisation, but
before commencing the development and implementation phase.

This chapter introduces an IS/IT readiness model which is capable of
measuring the organisational four key elements following the maturity modelling concept. The rationale behind the model is that for a new system to

Figure 10.1 Domains of readiness.

be successfully implemented in an organisation, the four key soft elements need to be at a level of readiness that is adequate for that particular system. For example, it is extremely unlikely that an organisation based on a heavy hierarchal management structure will successfully utilise Internet-based management systems which are designed to ease the communication and the decision-making process between the different stakeholders (see Chapter 2, Case study 2.4). Also, an organisation that tries to overcome the large backlog and heavy maintenance load of its IS/IT is unlikely to be able to develop substantial strategic information systems.

## 10.2 The development of an IS/IT readiness model

The proposed readiness model adopts the concept of maturity models to provide a base for measuring improvements in the four key soft elements. Maturity levels are applied to both the current organisational status and the target status required for the successful implementation of new IS projects. By identifying the current and the required organisational status, the readiness gap can be determined, and the route of progress on the maturity steps becomes visible. Progress can be accomplished only when organisations move through the levels in sequential order.

The model also adopts easy-to-use qualitative measures to ensure consistency and in-depth insight of the organisational situation. It provides a management tool for quick and easy identification of the organisational likely problem area(s) and suggests an improvement for a particular IS/IT project prior to its implementation phase.

In general the model provides an overall organisational assessment through predefined attributes for each of the four key elements. These attributes are mainly based on previous efforts in this field. For example, four of the components that were described in the Seven "Ss" used by McKinsey and Company (Pascale and Athos 1981) and also used in the Galliers and Sutherland's model (Galliers and Sutherland 1991) have been adopted in this model. These components are structure, systems, staff, and skill. The model also incorporates other influential factors for IS/IT success including culture, leadership style and position of IS/IT head. Since the functionalities of a specific information system can span over different departments in an organisation, the identified attributes provide equally viable measurable criteria for these business areas. If the IS/IT project is confined to and isolated in a specific business unit within an organisation, then the model is to be applied in that unit only.

Since the model is of a normative type with a wide focus to provide a pragmatic management diagnostic tool, the concept of several product and process normative IS/IT measurement models has also been adopted. The levels and descriptions of the components have been reviewed with the aim of extracting useful components for the model. Table 10.1 lists those models that have been

*Table 10.1* Models referenced for attributes extraction

| Referenced model | Features adopted |
| --- | --- |
| Capability Maturity Model (SW-CMM) (Paulk et al. 1993) | Maturity modelling and levelling, process elements; guidelines of progress |
| Capability Maturity Model (SE-CMM) | Process elements, 'generic practices' concept |
| • People-CMM (SEI 2001)<br>• Personal Software Process (PSP) (Humphrey 1995)<br>• Team Software Process (TSP) (Humphrey et al. 1999) | Elements in people and environment domains' attributes (skill, culture, leadership, staff); guidelines of progress |
| Trillium (Trillium 1996) | Model structure, Guideline (90% fulfilment), idea of being able to join process elements from CMM and ISO 9000 standards in one model |
| Bootstrap (Kuvaja et al. 1994) | Process elements, assessment method (multiple level assessment and overall average of level strength) |
| SPICE (Software Process Improvement And Capability Determination) (El Emam et al. 1998) | Process elements, concept of continuous verses level |
| Goal/Question/Metric (GQM) paradigm (Solingen and Berghout 1999) | A goal-centred alternative structure to the wide-focused normative for generalisation |
| ISO 9000 series | Guidelines (relationship between ISO and CMM), elements of attributes (process, culture, leadership) |
| • Nolan (1979)<br>• Bhabuta (1988)<br>• Earl (1989)<br>• Hirschheim et al. (1988)<br>• Galliers and Sutherland (1991) | Elements of attributes (skill, staff, leadership, head of IS/IT, systems, culture, structure) |
| Weill and Broadbent (1999) | Elements of attributes (systems, staff, skill) |

referenced and used for the attributes, model structure and guidelines of use. After components of each attribute have been formed and positioned in the proposed model's maturity levels, they were refined and the model was modified at different stages during its development. This modification process involved combining, deleting, rewording and/or changing level or attribute positions of different components. In the extraction process, the descriptions of each organisational element were chosen from the different normative models, categorised to be in their respective maturity levels. A wording, rewording and/

or combining process was then applied to the large pool of attributes to produce the levels of each of the four elements.

As illustrated in Figure 10.2, the attributes that define each of the four key
organisational elements are:

People – the main attributes are "staff", "skill" and "Head of IS/IT
function".

Process – the main attributes follow "general practices" as identified by the
Capability Maturity Model.

IT infrastructure – the main attribute are the current "systems" in terms of
hardware, communication, networks and software.

Work environment – the main attributes are "management style/leader-
ship", "structure" and "culture".

## 10.3 Description and use of the IS/IT
## readiness model

The model describes the readiness of the organisation for an IS/IT project in
terms of the four elements embracing eight attributes, as stated in Figure 10.3
(see also Salah and Alshawi 2005). Each of the attributes is described in six
levels where each represents a maturity level describing the organisational status
in terms of that particular attribute.

The model describes the four elements (people, process, IT infrastructure, and work environment) in detail using criteria describing the attributes
associated with each of them as they might occur in each of the six maturity
levels. Each of the criteria describing an attribute comprises aspects of how
the status of that particular attribute should be at different maturity levels.
Some of the descriptions of the attributes of the early levels might be understood to have a negative notion. This should not be the case. The model is
attempting to describe, depending on accumulated experiences and previous
research models, the status of organisations' IS/IT maturity at each of those
levels.

The following example, from the IT infrastructure elements, Level 1 (see
Section 10.4.1), illustrates how the attributes could be interpreted:

IS/IT systems that are developed/purchased at this maturity level tend to be
small and mainly off-the-shelf financial packages, where the decisions
regarding acquiring them tend to be of an ad hoc nature. Those systems
also tend to be independent of each other (stand-alone) and built/
purchased in isolation from other IS/IT located in the organisation or even
in the same group. Investment decisions regarding those systems are made
at low levels in the organisation, mainly at individual or group level, and

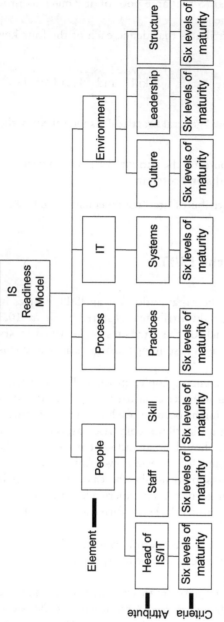

*Figure 10.2* Structure of the IS readiness model.

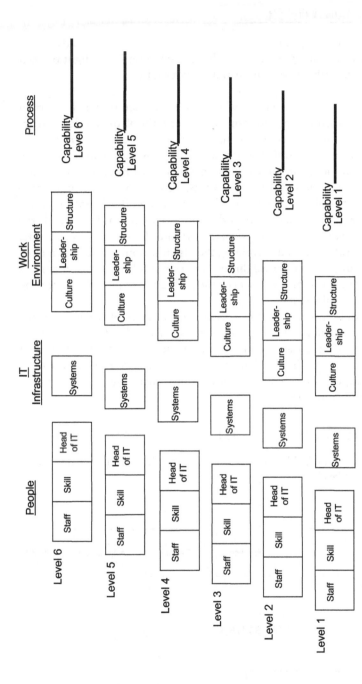

*Figure 10.3* The IS readiness model.

according to what the group's management see as the group's needs. Those needs and how to fulfil them are mainly decided upon in the light of what management see other people have and do externally.

The model is of a general nature where users can determine the current capability of their organisations in terms of the criteria of the model's attributes. The status of an organisation is measured against the criteria of each attribute. If the status of the organisation corresponds with the overall description of two maturity levels in almost equal manner, then both maturity levels could be said to match the status of the organisation for that particular attribute. If the organisational status corresponds with the overall description of two levels in almost equal manner, then both levels could be said to match the organisational situation for that particular attribute. It is important to state here that the methods used in determining the organisational status are mainly qualitative by nature, i.e. determined through interviews, observation, document review, etc., and thus the more experienced the evaluation team is the better and more accurate the results will be.

The model should also be used in specifying the levels of maturity of the different attributes mandated by the pre-selected IS. Those levels are considered to be the target levels for the organisation to achieve in order to implement successfully an IS project. The difference between the current organisational status and the target in terms of all the elements' attributes constitutes the readiness gap (see Figure 10.4).

## 10.4 The IS/IT readiness model

This section presents the IS/IT readiness model in the four elements along with their attributes and levels.

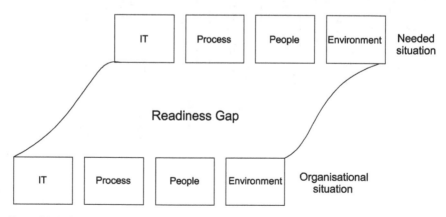

*Figure 10.4* Organisational IS/IT readiness gap.

### 10.4.1 IT infrastructure

*Systems*

LEVEL 1

- Almost all existing systems are small packages for financial operations.
- IS/IT development is ad hoc, where each unit invests independently from the rest of the organisation and the approval process of IS/IT project differs between units.
- Information systems are independent and unconnected organisation-wide or even within the same group, which makes IS/IT portfolios of each group differ from the rest of the organisation.
- IS/IT development, maintenance, implementation and training are made at the group level, where groups manage their own IS/IT resources according to each group's needs in isolation from the rest of the organisation.

LEVEL 2

- An increase in the number of IS/IT application systems is being developed or purchased but concentration is still on operational systems in the financial area while a small number of other core business-oriented systems are being developed.
- Many of the IS/IT application systems still overlap in purpose, function and data stored in the organisation, where only some hardware, system software, and possibly a network are shared between groups.
- A large maintenance load is being placed on IS/IT function because of the ad hoc nature of most systems.
- All data are stored in units' systems, except data needed for organisational reporting which are transferred to central systems.

LEVEL 3

- In-house IS/IT applications covering most major operation areas with office automation exist, but in an isolated stand-alone manner.
- Technical infrastructure consists of unconnected systems where no shared applications exist.
- Systems have been implemented in most operational areas in the organisation where the use of IS/IT services varies among the business units.
- Some IS/IT application systems have still been put together by users, and old user-developed systems are being used in an uncontrolled, uncoordinated manner even though new systems are centrally developed, installed, and operated by IS/IT function.

LEVEL 4

- All needed operational IS/IT is mostly in place and some DSS start to appear.
- Office automation is integrated and unified/standardised organisation-wide.
- An organisation-wide network exista, where all groups are connected and the central IS/IT function provides communication services for all groups in the organisation.
- The use of IS/IT is centrally coordinated throughout the organisation, where an effort is made by groups' IS/IT functions to follow standards set centrally.
- The organisation-wide network is starting to be utilised to connect users to whatever shared applications and information systems are needed.
- There is extensive use of standard e-mail messages throughout the organisation, and evidence of dependence on the organisation-wide network to conduct formal communication.

LEVEL 5

- Strategic IS/IT applications are developed with external-oriented data along with DSS and EIS. These are built over and based on the existing operational systems completed in the last stage.
- New systems rely heavily on gathering and processing external and internal data through the use of EDI systems with external entities such as customers, government and suppliers, which introduces problems of compatibility between external and internal data.
- New systems provide strategic advantages to the organisation or units, where IS/IT starts to be used to add value to organisational products and/or services.
- Most new systems are decentralised, but with central coordination and control.
- DSS and EIS systems are developed for the use of senior management.

LEVEL 6

- Inter-organisational systems exist with outside entities (government, suppliers, customers, etc.), with the use of Internet and e-commerce technology.
- There are shared inter-organisational systems (with suppliers, customers, government, etc.) using shared IS/IT infrastructure services such as the Internet.
- Intranet provision improves effectiveness.
- There are no geographic constraints on the provision of information.
- There exists a diverse hardware architecture according to each unit's needs.
- GSS and KMS systems are developed and successfully used.

## 10.4.2 People

### Staff

LEVEL 1

- There are no dedicated IS/IT staff or a small number of low level technicians and programmers.
- No manager is allocated responsibility for IS/IT.
- External contractors may be used to develop/install systems as required.
- New user recruits are not expected to have IS/IT-related skill.

LEVEL 2

- The small IS/IT staff consists, in addition to programmers and low level technicians, of system analysts where qualified individuals (mainly programmers and analysts) are selected, recruited, and transitioned into assignments.
- A DP manager, who has recently been appointed, is responsible for IS/IT function.
- IS/IT staff are now charged with the responsibility of adequately understanding the user requirements needed for systems' development.
- New user recruits are expected to have basic IS/IT skills.

LEVEL 3

- In addition to the programmers and analysts, dedicated IS/IT planners and database administrators are appointed.
- Almost all needed technical specialist staff are in-house.
- A technically oriented IS/IT manager is appointed, or the DP manager might have a change in title.
- The IS/IT workforce are coordinated with current and future IS/IT needs at both the organisational and unit levels.
- New user recruits are expected to have specific IS/IT-related skills.

LEVEL 4

- In addition to the programmers, systems analysts, and database administrators, business analysts now exist to act as a liaison between their units and the IS/IT function.
- A high level manager for IS/IT services area is appointed, with middle management status.
- Organisational staff (IS/IT and user) performance is quantified and measured against quantitative performance baselines.

LEVEL 5

- Core hybrid staff are sought, developed, and retained, while in some large organisations, some developed expertise is outsourced.
- Combining the roles of IS/IT and business planners to plan the strategic IS/IT for individual groups and the organisation as a whole, where the business/IS/IT planners have an experience from working in/with both users and the IS/IT function making them cross-disciplinary.
- The IS/IT manager has senior management status.
- An innovator workforce emerges in the organisation.

LEVEL 6

- Core staff are retained, and there is widespread outsourcing of expertise in some large organisations.
- Information staff are shared between large allied companies.
- There is widespread user ownership.
- The IS/IT function head becomes a full member of the board of directors to play an active part in setting strategic direction, but not only as an advisor, since strategic plans need to have the required IS/IT element "cooked" in them from the beginning.
- The existence of innovator workforce is widespread and rooted in the organisation.

*Skills*

LEVEL 1

- Users find it hard to acquire the skills to use the IS/IT that exists, and skills are individually based and jealously guarded from others.
- The emphasis is on technology rather than organisational or informational issues; there are limited technical skills in the organisation/unit.
- IS/IT skills are specific to individual IS/IT applications.
- Needed skills are of a low level, technical nature and there may exist limited advanced (programming or systems analysis) skills in the organisation.
- There is almost no IS/IT training provided by the organisation.

LEVEL 2

- Users begin to have the needed training and skills to use the new IS/IT that is being developed/purchased.
- There is still little in-house technical expertise in IS/IT development (methodology, structured techniques) and other important skills.
- IS/IT staff acquire the skills needed to develop and maintain complete

systems such as programming and analysis, in addition to being able to install off-the-shelf, ready-made packages.

- Project management skills are limited.
- IS/IT individuals have the skills required to perform their assignments and begin to have the relevant training and development opportunities.
- Individuals in the organisational workforce have remuneration and benefits based on their contribution and value to the organisation.

## LEVEL 3

- There is considerable technical competence in the organisation because of the well-developed IS/IT related skills (programming, analysis, security, networking etc.).
- The organisational workforce are constantly enhancing their IS/IT capabilities to perform their assigned tasks and responsibilities.
- The DP/IS/IT manager and staff lack – but work on – building interpersonal skills.
- Project management is realised to be needed at this stage, resulting in well-developed project management skills.

## LEVEL 4

- Systems staff have business skills, since knowledge and skills are required now of IS/IT staff beside their technical capability to fit in more with the rest of the organisation. At the same time users gain proper insight into IS/IT related issues.
- The IS/IT head and staff have good interpersonal skills.
- All individuals are involved in capturing/documenting their knowledge and experience from performing IS/IT-related work to be used in enhancing their competency and performance.
- The organisation's workforce have the ability to mentor, i.e. to use the IS/IT-related experience to provide personal support and guidance and to share professional and personal skills and experiences with less experienced staff, with the goal of developing these individuals. This guidance can involve developing knowledge, skills, and process abilities, improving performance, handling difficult situations, and making career decisions.

## LEVEL 5

- Core technical skills are developed, and some expertise might be outsourced.
- IS/business planners have the skill and experience to plan the strategic information systems for individual units and the organisation as a whole, where this experience has been gained from working in/with both users and IS/IT function.

- As the IS/IT function becomes an integral part of the organisation, hybrid skills are used wherever possible, and entrepreneurial skills start to be encouraged within the IS/IT and user workforce; very knowledgeable users of IS/IT become quite normal, and they now contribute freely to the whole IS/IT operation without any sensitivity from IS/IT function personnel.
- The IS/IT head has senior executive skills.
- Individuals and workgroups are continuously improving their IS/IT-related capability for performing their work processes.

## LEVEL 6

- The Head of IS/IT, also a member of the board of directors, has all the skills and knowledge necessary for both IS/IT and business roles.
- Consensus management skills are widely evident.
- Skills are shared between allied large organisations, and outsourcing of specialist skills is widespread.
- IS/IT staff are keeping up with the strategic needs of the groups they work with.
- Individuals and workgroups are optimising their IS/IT-related capability for performing their work processes.

## Head of IS/IT function

### LEVEL 1

- There is no individual responsible for systems.

### LEVEL 2

- There is a DP manager under a financial control group.

### LEVEL 3

- There is a new IS/IT manager appointed – or the DP manager might have a change in title – who is viewed as a technocrat.

### LEVEL 4

- There is an IS/IT manager who has middle manager status.

### LEVEL 5

- There is an IS/IT manager who has senior manager status.

**LEVEL 6**

- There is an IS/IT manager who is on the Board of Directors.

### 10.4.3 Work environment

*Management style/leadership involvement*

**LEVEL 1**

- There is little concern for the potential utility of IS/IT.

**LEVEL 2**

- IS/IT is considered to be the concern of technologists not management; there is support for IS/IT, although priority and thrust are to minimise the expense of IS/IT utilisation.

**LEVEL 3**

- IS/IT is considered to be one of the many ways to cut costs in the firm and the expenditure on IS/IT as a way of saving cost; IS/IT is looked at as a utility that provides service at minimum cost.

**LEVEL 4**

- IS/IT is seen to be vital for the smooth functioning of operations.

**LEVEL 5**

- IS/IT is considered as one of the vital parts of the competitive strategy, where a flexible IT infrastructure is perceived as an asset for competitive edge and is brought up in this way during project justification.

**LEVEL 6**

- IS/IT is seen as the single most critical factor for the organisation.

### Structure

**LEVEL 1**

- There is no central IS/IT function or no formal IT/IS organisational structure because the organisation does not consider the actual benefit of use of IS/IT.

- Responsibility for IS/IT is dispersed throughout the organisation.
- Head of the organisation or site CEO is actively involved in IS/IT purchasing.
- Business units are independent with little synchronisation in regard to IS/IT.

### LEVEL 2

- A separate DP/IS/IT function has recently been introduced, and groups are encouraged to seek advice from this newly formed central IS/IT function.
- There is a decentralised responsibility for IS/IT function, where groups have full freedom in managing their IS/IT with an increased self-reliance regarding IS/IT matters that is apparent throughout the organisation.

### LEVEL 3

- Official power is vested in the IS/IT function, where a new technical IS/IT manager is appointed to IS/IT manager (or the DP manager might have a change in title), and this coincides with a similar change in department name.
- Organisation-wide IS/IT architecture policy and standards for telecommunications, preferred suppliers, e-mail, etc., exist.
- Management of the IS/IT function is centralised.
- IS/IT staff seek control of IS/IT matters.

### LEVEL 4

- IS/IT function is well established and its mission is to exploit the IS/IT for business purposes and provide competitive IS/IT in a partnership environment with users.
- There is decentralised responsibility for IS/IT services with central standards and policy for coordination, implementation and utility.
- The units' IS/IT function reports to units' business manager.
- There is a significant degree of involvement of users in IS/IT-related decisions, and IS/IT investments are derived from users' stated needs.

### LEVEL 5

- There is a federal decentralised management structure with flexibility to support IS/IT initiatives, and there exist a strategic coalition and partnership between IS/IT function and user groups in large organisations.
- There are decentralised IS/IT function units, with a central IS/IT function providing an organisation-wide communication system, major data processing, and large-scale hardware in large organisations.
- In some organisations the budget of central IS/IT function is paid for by business units for services rendered.

LEVEL 6

- There is central co-ordination of the strategic coalition between IS/IT function and user groups.
- The still decentralised IS/IT function units, with a central IS/IT function, provide organisation-wide communication system, major data processing, and large- scale hardware.
- There is an overall integration of views regarding IS/IT in the organisation.

## Culture

LEVEL 1

- The relationship between the user and the technical staff, that may exist in the organisation or be contracted from outside, is one of support for the existing IS/IT products.
- There is no recognition in the organisation of the importance of working towards building a constructive relationship between IS/IT function and users or among staff members in the same group.

LEVEL 2

- IS/IT function wants to satisfy user needs but there is no control of user IS/IT-related activities.
- The organisation starts to recognises the need for building the capabilities for implementing timely communication across the organisation and for the workforce to acquire the skills for sharing information and coordinating their activities efficiently.

LEVEL 3

- IS/IT function seeks control over the activities of users concerning IS/IT matters; this is faced by implicit and explicit resistance.
- There exists a flow of information within the organisation, where voices start to be raised for using this flow of information to incorporate the knowledge of individuals into decision-making processes, and to gain their support to establish work commitments on both IS/IT function and user sides.

LEVEL 4

- IS/IT function supports the activities of users.
- There exists an emphasis on organisational integration between work-groups, among which is the IS/IT function.

- Workgroups (IS/IT and users) have the responsibility for and authority to determine how to conduct their business activities most effectively.
- An improvement in the efficiency and quality of interdependent work results from the integration of the capabilities and knowledge of different workgroups on both IS/IT function and user sides, and with each other.

### LEVEL 5

- IS/IT function and users cooperate on equal bases as partners.
- In addition to the existence of the characteristics stated in the second, third, and fourth points of Level 4 above, there exists a continuous striving for integration of organisational workgroups.

### LEVEL 6

- IS/IT function is a central resource which has mainly a coordinating role.
- There is a desire to optimise integration of organisational workgroups.

### 10.4.4 Process

*Generic practices*

### LEVEL 1

- Ad hoc level

### LEVEL 2

GG 1  Achieve specific goals
      GP 1.1     Identify work scope
      GP 1.2     Perform base practices

### LEVEL 3

GG 2  Institutionalise managed process
      GP 2.1     Establish an organisational policy
      GP 2.2     Plan the process
            1  Obtain management sponsorship for performing the process.
            2  Define and document the process description.
            3  Define and document the plan for performing the process.
            4  Review the plan with relevant stakeholders and get their agreement.

          5   Revise the plan as necessary.

GP 2.3     Provide resources

GP 2.4     Assign responsibility

1   Assign overall responsibility and authority for performing the process.
2   Assign responsibility for performing the specific tasks of the process.
3   Confirm that the people assigned to the responsibilities and authorities understand and accept them.

GP 2.5     Train people

GP 2.6     Manage configurations

GP 2.7     Identify and involve relevant stakeholders

1   Identify stakeholders relevant to this process and decide what type of involvement should be practised.
2   Share these identifications with project planners or other planners as appropriate.
3   Get stakeholders involved as planned.

GP 2.8     Monitor and control the process

1   Measure actual performance against the plan.
2   Review accomplishments and results of the process against the plan.
3   Review activities, status, and results of the process with the immediate level of management responsible for the process and identify issues.
4   Identify and evaluate the effects of significant deviations from the plan.
5   Identify problems in the process and in the plan.
6   Take corrective action when requirements and objectives are not being satisfied, when issues are identified, or when progress differs significantly from the plan.
7   Track corrective action to closure.

GP 2.9     Objectively evaluate adherence

GP 2.10    Review status with higher-level management

## LEVEL 4

GG 3  Institutionalise defined process

GP 3.1     Establish a defined process

1   Select the standard process that best fits the specific instances from the organisation's set of standard processes.
2   Establish the defined process by tailoring the selected standard processes and other process assets according to the organisation's tailoring guidelines.

3   Ensure that the organisation's process objectives are appropriately addressed in the defined process.

4   Document the defined process and the records of the tailoring.

5   Revise the description of the defined process as necessary.

GP 3.2   Collect improvement information

1   Store process and product measures in the organisational measurement repository.

2   Submit documentation for inclusion in the organisational library of process-related assets.

3   Document lessons-learned from the process for inclusion in the organisational library of process-related assets.

4   Propose improvements to the organisation's process assets.

## LEVEL 5

GG 4   Institutionalise quantitatively managed process

GP 4.1   Establish quality objectives

1   Obtain quantitative objectives for the project's defined process or, if they are not available, from other sources.

2   Allocate the quantitative objectives to the process.

GP 4.2   Stabilise sub-process performance

1   Statistically manage the performance of one or more sub-processes that are critical contributors to the overall performance of the process.

2   Estimate the ability of the process to achieve its established quantitative objectives considering the performance of the statistically managed sub-processes.

3   Incorporate selected process performance measurements into the organisation's process performance baselines.

## LEVEL 6

GG 5   Institutionalise an optimising process

GP 5.1   Ensure continuous process improvement

1   Establish and maintain quantitative process improvement objectives that support the organisation's business objectives.

2   Identify process improvements that would result in measurable improvements to process performance.

3   Define strategies and manage deployment of selected process improvements based on the quantified expected

benefits, the estimated costs and impacts, and the measured change to process performance.

GP 5.2    Correct common cause of problems

## 10.5 Case study 10.1: implementation of the readiness model

This section will follow up the two detailed case studies that were described in Chapter 4: ServInst and OilCo. The IS readiness model was applied to each of the two cases with the aim of determining the readiness gap of the two organisations. Although the readiness model was applied at almost the end of the development and implementation phase of these cases, nevertheless it adequately reflected the organisation's readiness level in the four elements. The following sections will discuss the two cases; the status of each organisation will be examined prior to the start of the project and at the end of the project to then determine the expected target level.

### 10.5.1 Case study 4.1: ServInst

After the analysis of this case study (see Chapter 4, Section 4.2.3) the IS/IT readiness model was applied to the four elements.

### People: staff

#### STATUS PRIOR TO PROJECT

At the start of the project, the readiness level mostly agreed with the general description of Level 3. This is because the organisation had programmers, analysts and a database administrator for the personnel system, and almost all needed technical staff. Also the IS/IT staff recognised the current and future IS/IT needs at both the organisational and unit levels and new user recruits were expected to have specific IS/IT-related skill. Additionally the technically oriented IS/IT manager had a middle management title.

#### STATUS AT TERMINATION OF PROJECT

The status of the staff at the termination of the project deteriorated to mostly agree with the general descriptions of some of Level 1 and Level 2 criteria. This is because project team members had changed from both the organisation and vendor sides to less experienced junior staff. This had an impact on the overall staff status.

## TARGET STATUS

The proposed target situation needs to agree with the general description of Level 5 because the system spans over all organisational functions and needs an IS/IT head that has at least the power of user managers. Also an organisation that depends to a great extent on IS/IT and has been in such a situation for 20 years needs to have a strategic vision of IS/IT along with staff who can plan accordingly. This also requires staff on both the user and IS/IT sides who would understand each other's work to be able to have integration of groups and systems. The organisation needs to have a core hybrid staff to be developed and retained. It also needs to combine the roles of IS/IT and business planners to plan the strategic IS/IT for individual groups and for the organisation as a whole, where the business/IS/IT planners acquire experience from working with both users and the IS/IT unit (multidisciplinary). The organisation also needs to restore to the IS/IT manager the senior management status he had a decade ago.

The result of the Staff Analysis is as follows:

| People Staff | | | | |
|---|---|---|---|---|
| Status | Prior to | At termination | Target | Gap |
| Maturity level | 3 | 1–2 | 5 | 3, 4, 5 |

### People: skill

## STATUS PRIOR TO PROJECT

The status mostly agreed with the general description of Level 2 because users in the organisation had the required training and skills to use the IS/IT. The technical expertise was limited in scope where IS/IT project development was mainly done to accommodate new changes in governmental laws and regulations that reflect into the way ServInst conducts its work. This made project management skills limited to such small projects, but required the organisation to provide IS/IT staff who had the relevant training and development opportunities needed to perform such assignments. Since the IS/IT job market has a high turnover, the organisation felt obliged to provide its IS/IT staff with remuneration and benefits based on their contribution and value to the organisation in order to retain them.

## STATUS AT TERMINATION OF PROJECT

The status of the skill attribute deteriorated at the project termination to mostly agree with the general descriptions of some of Level 1 criteria. Project team members had changed from both the organisation and vendor sides to less experienced junior staff.

TARGET STATUS

The proposed target situation needs at least to comply with with the general description of Level 4, in addition to the strategic planning capabilities and senior management skills in Level 5. This is because the system spans over all organisational functions, and therefore requires an IS/IT head with senior management skills. Also an organisation that depends to a great extent on IS/IT needs to have a strategic view of IS/IT, and a staff that can plan accordingly. This also requires good communication skills to enable the integration of the different organisational groups.

The result of the Skill Analysis is as follows:

| People Skill | | | | |
|---|---|---|---|---|
| Status | Prior to | At termination | Target | Gap |
| Maturity level | 2 | 1 | 4–5 | 2, 3, 4, 5 |

*People: Head of IS/IT*

STATUS PRIOR TO PROJECT

At the beginning of the project, the situation mostly agreed with the general description of Level 4, where the IS/IT manager has middle manager status.

STATUS AT TERMINATION OF PROJECT

The situation of the Head of IS/IT at the termination of the project remained the same as it was at the beginning, at Level 4.

TARGET STATUS

The proposed target situation needs to agree with the general description of Level 5, so that the IS/IT head acquires the power needed to manage and enforce policies of IS/IT that are organisation-wide and to be able to participate in strategic decisions.

The result of the Head of IS/IT Analysis is as follows:

| People Head of IS/IT | | | | |
|---|---|---|---|---|
| Status | Prior to | At termination | Target | Gap |
| Maturity level | 4 | 4 | 5 | 5 |

## Work environment: leadership

### STATUS PRIOR TO THE PROJECT

At the beginning of the project, the status mostly agreed with the general description of Levels 2 and 3 because the attitude of the top management fluctuated according to the external and internal pressures. It is important to note that the cost issue was of little importance for top management in the organisation because it was a governmental institute that had to hand a large amount of income from government and other institutions and a large amount of return on its large investments. Level 3 leadership in a government-run organisation would consider IS/IT mainly as a utility to provide services.

### STATUS AT TERMINATION OF PROJECT

The status of the leadership at the termination of the project agreed mainly with the Level 2 description. Top management lost enthusiasm and left the project team with little support.

### TARGET STATUS

The proposed target situation agrees most with the general description of Level 5's strategic view of IS/IT, because the organisation is an important national service provider that depends almost entirely on IS/IT in running its work processes. In the case of lack of success, a national problem might occur.

The result of Leadership Analysis is as follows:

| Environment Leadership | | | | |
|---|---|---|---|---|
| Status | Prior to | At termination | Target | Gap |
| Maturity level | 2–3 | 2 | 5 | 3, 4, 5 |

## Work environment: culture

### STATUS PRIOR TO PROJECT

At the beginning of the project, the status mostly agreed with the general description of Level 1 because the relationship between users and the technical staff was one of support and maintenance for the existing IS/IT, and there was no recognition in the organisation of the importance of working towards building a constructive relationship between IS/IT unit and users or among staff members.

### STATUS AT TERMINATION OF PROJECT

The status of the culture at the termination of the project remained the same as it was at the beginning, i.e. on Level 1.

### TARGET STATUS

The proposed target situation should at least agree with the general description of Level 4 and preferably Level 5, because the system is a highly integrated system that cuts through unit and group boundaries. In such an environment, the relationships between all organisation's workforce need to be those of integration and cooperation. The organisation needs the IS/IT unit to support the activities of users, and both cooperate on equal bases as partners. There also needs to exist an emphasis on organisational integration between different workgroup capabilities and knowledge on both IS/IT unit and user sides. Those workgroups (IS/IT and users) need to have the responsibility and authority for determining how to conduct their business activities most effectively.

The result of Culture Analysis is as follows:

| Environment  Culture | | | | |
|---|---|---|---|---|
| Status | Prior to | At termination | Target | Gap |
| Maturity level | I | I | 4–5 | 2, 3, 4, 5 |

## Work environment: structure

### STATUS PRIOR TO PROJECT

At the beginning of the project, the status mostly agreed with the general description of Level 3, because the IS/IT unit's technically oriented manager had a middle management authority and status, and management of the IS/IT unit was centralised and sought control of IS/IT matters in the organisation. Also there were no organisational IS/IT architecture policy and standards; rather, there were accepted lists for choices of preferred suppliers of IS/IT products.

## STATUS AT TERMINATION OF THE PROJECT

The status of the structure at the termination of the project remained the same as it was at the beginning, at Level 3.

## TARGET STATUS

The proposed target situation needed to agree with the general description of Level 4 as concerns the need to have architecture policy and standards for coordination, implementation and utility of IS/IT, but also to emphasise (as in Level 5) that the IS/IT unit and project team need to have senior authority to execute their mission to implement a strategic coalition and partnership with user groups based on a strategic view of IS/IT.

The result of Structure Analysis is as follows:

| Environment Structure | | | | |
|---|---|---|---|---|
| Status | Prior to | At termination | Target | Gap |
| Maturity level | 3 | 3 | 4–5 | 4, 5 |

## IT infrastructure: systems

## STATUS PRIOR TO PROJECT

At the beginning of the project, the status mostly agreed with the general description of Level 4 but in an unfulfilled way. This was because all needed operational IS/IT was mostly in place but did not fully serve the organisational needs. Office automation existed, but in an isolated not standardised manner. Also ther eexisted an organisation-wide network, where all groups were connected and the central IS/IT unit provided communication services for all groups, although it did not accommodate the planned expansion and did not support e-mail.

## STATUS AT TERMINATION OF THE PROJECT

The status of the systems at the termination of the project remained the same as it was at the beginning, at a weak Level 4.

TARGET STATUS

The proposed target situation needs to still agree with the general description of Level 4 but with a new network infrastructure that utilises e-mail and connects to the Internet to facilitate better service for the external customers.

The result of the Systems Analysis is as follows:

| IT Systems | | | | |
|---|---|---|---|---|
| Status | Prior to | At termination | Target | Gap |
| Maturity level | 4 | 4 | 4 | Improvements |

*Process: practices*

STATUS PRIOR TO PROJECT

At the beginning of the project, the status mostly agreed with the general description of level 2-GG 1.

STATUS AT TERMINATION OF PROJECT

The status of the process at the termination of the project had improved to include some of the generic practices of level GG 2. Those additional practices were the result of the "strategic study" conducted after the start of the project and are designated GP 2.1, GP 2.2, and GP 2.7.

TARGET STATUS

According to the project manager (who was interviewed specifically for this purpose) the proposed target situation needs to agree with the general description of Level 4-GG 3.

The result of the Process Practices Analysis is as follows:

| Process Practices | | | | |
|---|---|---|---|---|
| Status | Prior to | At termination | Target | Gap |
| Maturity level | 2-GG 1 | 3-GG 2 | 4-GG 3 | 3-GG 2 |

### 10.5.2 Case study 4.2: OilCo

After the analysis of this case study (see Chapter 4, Section 4.3.5) the IS readiness model was applied to the four key elements.

#### People: staff

##### STATUS PRIOR TO PROJECT

At the beginning of the project, the status mostly agreed with the general description of Level 2: the available IS/IT staff for the project consisted, in addition to a few programmers and low level technicians, of a few system analysts, and the recently appointed group head/manager. The small IS/IT staff consisted, in addition to the programmers and low level technicians, of system analysts, where qualified individuals (mainly programmers and analysts) were selected, recruited and transitioned into assignments:

- A DP manager who was appointed was responsible for IS/IT function
- IS/IT staff were charged with the responsibility of adequately understanding the user requirements needed for systems' development
- New user recruits were expected to have basic IS/IT skill

##### STATUS AT TERMINATION OF PROJECT

The status of the staff at the end of the project mostly agrees with the general description of some Level 3 criteria: in addition to the programmers and analysts, there were dedicated database administrators who were appointed during the project. Almost all needed technical specialist staff were in the team and the team head was promoted to manager status. The IS/IT staff were coordinated with current and future IS/IT needs at both the organisational and unit levels in regard to the new system, and users were expected to have specific IS/IT-related skills to use the system.

##### TARGET STATUS

The current situation (Level 3) seems to fulfil the proposed target situation concerning the staff attribute. The group seems to have enough personnel for maintaining and updating the system. The corrective steps taken by top management, and the project and deployment team leaders were fruitful in providing the teams with the needed staff to fit a Level 3 description. Providing the budget for hiring the needed staff, allowing qualified staff to join the team from other groups in the company, and promoting the team head to a middle management status, all contributed to the achievement of the target staff level.

The result of Staff Analysis is as follows:

| People Staff | | | | |
|---|---|---|---|---|
| Status | Prior to | At end | Target | Gap |
| Maturity level | 2 | 3 | 3 | – |

*People: skill*

STATUS PRIOR TO PROJECT

At the beginning of the project, the status of skill attribute of both the users and IS/IT people involved in the project mostly agreed with the general description of Level 1: staff did not have the required skills or the knowledge of the organisational needs as required for the project. These staff were allocated to the project from different parts of the organisation with no specific, related skills.

STATUS AT TERMINATION OF PROJECT

At the end of the project, there was a difference between the skill of the IS/IT staff and users. This occurred because of the support the IS/IT people received from top management, while users still suffered from resentment for the project, hindering their participation in the training programmes and their gaining experience from using the system. The users' skill improved in comparison to before the project, but not as much as needed. The skill situation of the user staff mostly agrees with some of the general descriptions of Level 2.

The level of skill of the IS/IT staff at the end of the project mostly agrees with some of the general descriptions of the Level 3 criteria: considerable technical competence existed among IS/IT staff after being trained and acquiring experience from working in the project. Those skills, which included programming, analysis, security, networking, etc., were enhanced by the experience and internal seminars held within the group. Because of being faced by resentment from users, the project group recognised the need to acquire interpersonal skills and develop their project management skills, which were the subject of some of the short seminars held within the group.

TARGET STATUS

The current situation (Level 3) seems to fulfil the needed target situation concerning skill of IS/IT staff. They have the needed skill for operating, maintaining and updating the system. The corrective steps taken by the project leader and the deployment team's head by providing training to the teams' staff while in the project, along with the support received from new top management provided the team with the needed skills, either through new staff recruitments or training the existing ones.

The user skills need to be improved to reach Level 3 which seems to be the adequate level for such a system that requires the skills for using it. Users need to work constantly on enhancing their IS/IT capabilities to perform their assigned tasks and responsibilities.

The result of Skill Analysis is as follows:

| People Skill | | | | |
|---|---|---|---|---|
| Status | Prior to | At end | Target | Gap |
| Maturity level | I | 2(user), 3(IS/IT) | 3 | Level 3 (user) |

*People: Head of IS/IT*

STATUS PRIOR TO PROJECT

At the start of the project, the situation mostly agreed with the general description of Level 2, i.e. the DP manager was under a financial control group/section.

STATUS AT TERMINATION OF PROJECT

At the end of the project the Head of DP became a manager with a job of a technical nature, agreeing with the description in Level 3.

TARGET STATUS

The proposed target situation for the DM head is to be of middle manager status, but not viewed as a technical manger, to be able to participate in the business-related decisions where s/he can bring into them a technological perspective. At the time, top management views and attitudes are moving in this direction. The target situation mostly agrees with the description of the Level 4 characteristics.

The result of Head of IS/IT Analysis is as follows:

| People<br>Head of IS/IT | | | | |
|---|---|---|---|---|
| Status | Prior to | At end | Target | Gap |
| Maturity level | 2 | 3 | 4 | Level 4 |

*Work environment: leadership*

STATUS PRIOR TO PROJECT

At the beginning of the project the status mostly agreed with the general description of Level 1, i.e. leadership had little concern for the potential utility of IS/IT.

STATUS AT TERMINATION OF PROJECT

The status of the leadership at the end of the project was improved dramatically by the change of top management to agree mainly with Level 4 criteria. The new management considered IS/IT to be vital for the smooth functioning of operations.

TARGET STATUS

For a system so vital to a company in the oil sector, the integrity and ease of accessibility to data of such importance requires a management that, at least, considers IS/IT as crucial for the company's operations. This requires providing the necessary budget and support to an IS/IT project as important as the DM project. The current leadership level is adequate as a target level.

The result of Leadership Analysis is as follows:

| Work Environment<br>Leadership | | | | |
|---|---|---|---|---|
| Status | Prior to | At end | Target | Gap |
| Maturity level | 1 | 4 | 4 | – |

## Work environment: culture

### STATUS PRIOR TO THE PROJECT

The culture attribute situation from the beginning of the project until just before the end was mostly in agreement with the general description of Level 1. This situation was one of lack of cooperation between the development team and users.

### STATUS AT TERMINATION OF PROJECT

The status of the culture attribute at the end had improved after the introduction of the user awareness programme and allocating the support personnel on-site with the users. The current situation agrees with the general theme of Level 2 culture criteria in the readiness model.

### TARGET STATUS

The organisation needed to improve communication and implement a program to take it to meet the criteria in Level 4 for the system to be fully successful. The system spans over many groups in a complementary workflow fashion. There is a need for integration between workgroups, among which is the IS/IT unit, for the system to be fully successful and improve the efficiency and quality of interdependent work. This proposed target situation mostly agrees with Level 4 criteria of culture in the readiness model. Before the groups reach this level, IS/IT needs to have control of activities regarding the workflow of the system. Without such control and without making users aware of the benefits, they will continue to ignore the system and keep doing work according to their old ways. This intermediary level mostly follows the general description of Level 3 criteria.

The result of Culture Analysis is as follows:

| Work Environment Culture | | | | |
|---|---|---|---|---|
| Status | Prior to | At end | Target | Gap |
| Maturity level | 1 | 2 | 4 | 3,4 |

*Work environment: structure*

STATUS PRIOR TO PROJECT

At the start of the project the status mostly agreed with the general description of Level 2: the team had been assigned the responsibility for developing the system. Responsibility for IS/IT was decentralised among groups, where groups had full freedom in managing their IS/IT with an increased self-reliance regarding IS/IT matters that was apparent throughout the organisation.

STATUS AT TERMINATION OF PROJECT

The current structure attribute situation was altered at end of the project when the project team's position was elevated. At the time, the structure attribute situation started to agree with the general theme of Level 3 criteria.

TARGET STATUS

The current structure attribute situation would be adequate to achieve most of the project's aims. For the system to be successful as a base for other Data Management systems organisation-wide, the structure attribute situation needs to have the position of IS/IT unit on a higher level to be able to coordinate this bigger task. The proposed target situation mostly agrees with the level 4 criteria described in the readiness model.

The result of Structure Analysis is as follows:

| Work Environment Structure | | | | |
|---|---|---|---|---|
| Status | Prior to | At end | Target | Gap |
| Maturity level | 2 | 3 | 4 | 4 |

*IT infrastructure: systems*

STATUS PRIOR TO PROJECT

At the beginning of the project, the systems attribute status mostly agreed with the general description of Level 3: IS/IT applications covered most major operation areas and office automation existed but in an isolated, stand-alone manner. In different parts of the organisation, technical infrastructure consisted of unconnected systems, but systems had been implemented in most operational areas in the organisation where the use of IS/IT services varied among the business units. Also some IS/IT applications had been requested and implemented by users, and the old user-developed systems were being used in an uncontrolled, uncoordinated manner.

### STATUS AT TERMINATION OF PROJECT

The status of the systems attribute at the end of the project mostly agrees with the general description of Level 4 in the readiness model. Some criteria in this level have not yet appeared, but the aim was to implement them, such as DSS and the central standardisation of packages and office automation tools. There had been noticeable progress on the standardisation issue, but it had not been applied organisation-wide.

### TARGET STATUS

The proposed target situation still needs to mostly agree with the general description of Level 4 that seems to meet the situation required for the new system to function properly.

The result of Systems Analysis is as follows:

| IT Infrastructure Systems | | | | |
|---|---|---|---|---|
| Status | Prior to | At end | Target | Gap |
| Maturity level | 3 | 4 | 4 | – |

### Process: practices

### STATUS PRIOR TO PROJECT

After interviews with the OilCo project leader and Head of implementation team it was agreed that the level of the process attribute at the start of the project was a weak Level 3-GG2, i.e. Level 3-GG2 practices were fulfilled in a weak manner.

### STATUS AT TERMINATION OF PROJECT

The level for process attribute at the end of the project was Level 3-GG2, i.e. maintaining Level 3-GG2 but this time with practices being fulfilled most of the time.

## TARGET STATUS

Since the development and deployment teams were not experienced in conducting process improvement or BPR, the progress of the process attribute was slow.
It might be that the teams were improving the processes by "trial and error".
According to the project leader and deployment team head, level 4-GG3 would
be sufficient as the target process level. Even though it would be better to have
even more mature processes, it is unrealistic at this time to expect more than
this level from the organisation because of negative cultural attitudes towards
change. Collecting improvement information in Level 4-GG3 would be important for improving the business processes for a sector that needs to be able to
meet the projected increasing demand for oil production.

The result of Process Analysis is as follows:

| Process Practices | | | | |
|---|---|---|---|---|
| Status | Prior to | At end | Target | Gap |
| Maturity level | Weak 3-GG2 | 3-GG2 | 4-GG3 | 4-GG3 |

# References

Abudayyeh, O.Y. and Rasdorf, W.J. (1991) "Design of Construction Industry Information Management Systems", *Journal of Construction Engineering and Management*, ASCE 117(4): pp. 698–715.

Ahmad, I.U., Russell, J.S. and Abou-Zeid, A. (1995) "Information Technology (IT) and Integration in the Construction Industry", *Construction Management and Economics* 13(2): pp. 163–171.

Al-Ghani, S. (1998) "Definition of IT", available online at http://www.commerce.uq.edu.au/isworld/research/msg.02–12–1998.html (accessed February 2002).

Al-Mashari, M. (2001) "Process Orientation through Enterprise Resource Planning (ERP): A Review of Critical Issues", *Knowledge and Process Management* 8(3): pp. 175–185.

Al-Mashari, M. and Zairi, M. (1999) "BPR Implementation Process: An Analysis of Key Success and Failure Factors", *Business Process Management* 5(1): pp. 87–112.

Al-Mashari, M. and Zairi, M. (2000) "Revisiting BPR: A Holistic Review of Practices and Development", *Business Process Management Journal* 6(1): pp. 10–30.

Alshawi, M. (2000) *Integrated Construction Environments: Technology and Implementation*, in Proceedings INCITE 2000: Implementing IT to Obtain a Competitive Advantage in the 21st Century, HK Poly University, Hong Kong, pp. 36–53.

Alshawi, M. and Aouad, G. (1995) "A Framework for Integrating Business and Information Technology in Construction", *Civil Engineering Systems* (12): pp. 249–261.

Alshawi, M. and Hassan Z. (1999) "CONPLAN: Conceptual Data and Process Models for Construction Planning within an Integrated Environment", the *International Journal of Engineering, Construction and Architectural Management*, Blackwell Science, 6(2): pp 197–212.

Alshawi, M. and Ingirige, B. (2003) "Web Enabled Project Management: An Emerging Paradigm in Construction", *Automation in Construction* 12: pp. 349–364.

Alter, A. (1990) "The Corporate Make-over", *CIO* 4(3): pp. 32–42.

Alter, S. (1996) *Information Systems: A Management Perspective*, 2nd edn, Menlo Park, CA: Benjamin/Cummings.

Amit, R. and Schoemaker, P.J.H. (1993) "Strategic Assets and Organizational Rent", *Strategic Management Journal* 14: pp. 33–46.

Andersen, J., Baldwin, A., Betts, M., Carter, C., Hamilton, A., Stokes, E. and Thorpe, T. (2000) "A Framework for Measuring IT Innovation Benefits", *ITCON* 5: pp. 57–72.

Andresen, J.L. (2000) "The Unidentified Value of IT in the Construction Industry", in *Proceedings The International Conference on Construction Information Technology*, Hong Kong Polytechnic University, pp. 93–105.

Andrews, J. (1984) *Construction Project Management in Joint Ventures in Developing Countries*, *Unibeam*, journal of the National University of Singapore.

Andrews, K.R. (1987) *The Concept of Corporate Strategy*, IL: Irwin Inc.

Anthony, R.N. (1965) *Planning and Control: A Framework for Analysis*, Cambridge, MA: Harvard University Press.

Applegate, L. and Elam, K. (1992) "New Information Systems Leaders: A Changing Role in a Changing World", *MIS Quarterly* 16(4): pp. 469–490.

Arendt, C., Landis, R. and Meister, T. (1995) "The Human Side of Change – Part 4", *IIE Solutions*: pp. 22–27.

Argyris, C. (1977) "Double Loop Learning in Organisations", *Harvard Business Review* (September–October): pp. 115–125.

Argyris, C. (1992) *On Organisational Learning*, Oxford: Blackwell.

Argyris, C. and Schön, D.A. (1978) *Organisational Learning: A Theory of Action Perspective*, Boston, MA: Addison-Wesley.

Arif, M., Talhami, H. and Alshawi, M. (2006) "Implementation of an E-Government Initiative at Dreamland Municipality", *International Journal of Cases in E-Commerce* 2(2) available online at http://www.igi-pub.com/cases/details.asp?id=5681 (accessed 15 May 2007).

Atkin, B. (1987) "Computerising", *Chartered Quantity Surveyor* 9(7): p. 10.

Atkin, B.L. (2000) "Benchmarking the IT Performance of Construction Firms and Their Projects", in *Proceedings The International Conference on Construction Information Technology*, Hong Kong Polytechnic University, pp. 106–118.

Bailey, M. and Pearson, S. (1983) "Development of a Tool for Measuring and Analyzing Computer User Satisfaction", *Management Science* 29(5): pp. 530–545.

Ballantine, J., Bonner, M., Levy, M., Martin, A., Murno, I. and Powell, P. (1996) "The 3-D Model of Information Systems Success: the Search for the Dependent Variable Continues", *Information Resources Management* 9(4): pp. 5–23.

Barki, H. and Rivard, S. (1993) "Toward an Assessment of Software Development Risk", *Journal of Management Information Systems* 10(2): pp. 203–225.

Barney, J. (1991) "Firm Resources and Sustained Competitive Advantage", *Journal of Management* 17(1): 99–120.

Bate, R. (1995) *A Systems Engineering Capability Maturity Model, Version 1.1*, Software Engineering Institute, Carnegie Mellon University.

Becta (2005) "ICT Maturity", March, available online at http://www.becta.org.uk.

Betts, M. (1992) "How Strategic is our Use of Information Technology in the Construction Sector", *International Journal of Construction Information Technology* 1(1): pp. 79–97.

Bhabuta, L. (1988) "Sustaining Productivity and Competitiveness by Marshalling IT", in *Proceedings Information Technology Management for Productivity and Strategic Advantage*, IFIP TC-8 Open Conference, Singapore, pp. 371–377.

Bhatt, G.D. (2001) "Knowledge Management in Organisations: Examining the Interactions between Technologies and People", *Journal of Knowledge Management* 5(1): pp. 68–75.

Biggs, M. (1997) "Why Choose a Web-based Project Management Solution? (Buyers Guide)" *PC World* 15(10): p. 190.

Bjørn-Andersen, N. and Turner, J. (1994) *Creating the Twenty-first Century Organization: The Metamorphosis of Action* (Transforming Organisations with Information Technology), Amsterdam: Elsevier Science.

Bohn, R.E. (1994) "Measuring and Managing Technical Knowledge", *Sloan Management Review* 36(Fall): pp. 61–73.

Bonner, M. (1995) "DeLone and McLean Model for Judging Information System Success", in *Proceedings European Conference on IT Investment Evaluations*, Henley College, pp. 218–235.

Borins, S. (2002) "On the Frontiers of Electronic Governance: A Report on the United States and Canada", *International Review of Administrative Sciences*, 68: pp. 199–211.

Brancheau, J., Janz, B. and Wetherbee, J. (1996) "Key issues in information systems management: 1994–1995 SIM Delphi results", *MIS Quarterly* (June): pp. 225–242.

Breuer, J. and Fischer, M. (1994) "Managerial Aspects of Information Technology Strategies for A/E/C Firms", *Journal of Management Engineering* 10(4): pp. 52–59.

Broadbent, M. and Weill, P. (1997) "Management by Maxim: How Business and IT Managers Can Create IT Infrastructures", *Sloan Management Review* (Spring): pp. 77–92.

Brynjolfsson, E. and Hitt, L.M. (1998) "Beyond the Productivity Paradox: Computers are the Catalyst for Bigger Changes", *Communications of the ACM* 41(8): pp. 49–56.

Brynjolfsson, E. and Yang, S. (1996) "Information Technology and Productivity: A Review from Literature", *Advances in Computers* 43: pp. 179–214.

Building (2001) "E-commerce Crashes to Earth", *Building Magazine* 12(January).

Building Centre Trust (1999) *Electronic Exchange of Project Information: 3COM project, phase 2, Information Technology Construction Best Practice.* Case Study, Building Centre Trust.

Caldas, C.H. and Soibelman, L. (2003) "Automating Hierarchical Document Classification for Construction Management Information Systems", *Automation in Construction*, Elsevier, 12(4): pp 395–406.

Carneiro, A. (2000) "How Does Knowledge Management Influence Innovation and Competitiveness", *Journal of Knowledge Management* 4(2): pp. 87–98.

Carr, D. and Johansson, H.J. (1995) *Best Practices in Re-engineering: What Works and What Doesn't in the Re-engineering Process*, New York: McGraw-Hill.

Central Computer and Telecommunications Agency (1992) *Information Technology Infrastructure Library*, London: HMSO.

Cerullo, M. (1980) "Information Systems Success Factors", *Journal of Systems Management* 31(12): pp. 10–19.

Chandler, J. (1982) "A Multiple Criteria Approach for Evaluating IS", *MIS Quarterly* 6(1): pp. 61–74.

Chourides, P., Longbottom, D. and Murphy, W. (2003) "Excellence in Knowledge Management: An Empirical Study to Identify Critical Factors and Performance Measures", *Measuring Business Excellence* 7(2): pp. 29–45.

Ciborra, C. (1994) "The Grassroots of IT and Strategy", in C. Ciborra and T. Jelessi (eds) *Strategic Information Systems: A European Perspective*, Chichester: Wiley, pp. 3–24.

Ciborra, C. and Andreu, R. (1998) "Organisational Learning and Core Capabilities Development: The Role of IT", in R. Galliers and W. Baets (eds) *Information*

*Technology and Organisational Transformation: Innovation for the 21st Century Organisation*, Chichester: Wiley.

Clark, A.M., Atkin, B.L., Betts, M.P. and Smith, D.A. (1999) "Benchmarking the Use of IT to Support Supplier Management in Construction", *Electronic Journal of Information Technology in Construction* 4: pp. 1–16.

Clegg, C., Axtell, C., Damodaran, L., Farbey, B., Hull, R., Lloyd-Jones, R., Nicholls, J., Sell, R. and Tomlinson, C. (1997) "Information Technology: A Study of Performance and the Role of Human and Organizational factors", *Ergonomics* 40(9): pp. 851–871.

Clemons, E. (1995) "Using Scenario Analysis to Manage the Strategic Risks of Re-engineering", *Sloan Management Review* 36(4): pp. 61–71.

Cohen, M.D. and Bacdayan, P. (1994) "Organizational Routines are Stored as Procedural Memory: Evidence from a Laboratory", *Organization Science* (December): pp. 554–568.

Construct-IT (2000) *Introduction to SPICE*, Construct IT Centre of Excellence, University of Salford.

Cooper, R., Kagioglou, M., Aouad, G., Hinks, J., Sexton, M. and Sheath, D. (1998) "The Development of a Generic Design and Construction Process", in *Proceedings The European Conference on Product Data Technology*, Watford, UK.

Cooper, R. and Markus, M. (1995) "Human Engineering", *Sloan Management Review* 36(4): pp. 39–50.

CSC (1996) *New IS Leaders*, Computer Sciences Corporation Index Research, London, UK: Computer Sciences Corporation.

CSC Index (1994) *State of Re-engineering Report*, London: CSC Index.

Curren, M.T., Folkes, V.S. and Steckel, J.H. (1992) "Explanations for Successful and Unsuccessful Marketing Decisions: The Decision Maker's Perspective", *Journal of Marketing* 56(2): pp. 18–31.

Davenport, T. (1993) *Process Innovation: Re-engineering Work through Information Technology*, Harvard Business School Press. Boston, MA.

Davenport, T. (1998) "Putting the Enterprise into the Enterprise System", *Harvard Business Review* 76(4): pp. 121–131.

Davenport, T. (2002) *Knowledge Management Case Study, Knowledge Management at Hewlett-Packard*, Austin, TX.

Davenport, T. and Nohria, N. (1994) "Case Management and the Integration of Labor", *Sloan Management Review*: pp. 11–23.

Davenport, T. and Prusak, L. (1998) *Working Knowledge*, Boston, MA: Harvard Business Press.

Davenport, T. and Prusak, L. (2000) *Working Knowledge: How Organisations Manage What they Know*, Boston, MA: Harvard Business Press.

Davenport, T.H. and Short, J.E. (1990) "The New Industrial Engineering: Information Technology and Business Process Redesign", *Sloan Management Review*. 13(4): pp. 11–27.

Davis, F.D. (1993) "User Acceptance of Information Technology: System Characteristics, User Perceptions and Behavioral Impacts", *International Journal of Man-Machine* (38): pp. 475–487.

DeLone, W. and McLean, E. (1992) "Information System Success: The Quest for the Dependent Variable", *Information Systems Research* 3(1): pp. 60–95.

Deng, Z.M., Li, H., Tam, C.M., Shen, Q.P. and Love, P.E.D. (2001) "An Application of the Internet-based Project Management System", *Automation in Construction* 10: pp. 239–246.

Dhillon, G. and Backhouse, J. (1996) "Risks in the Use of Information Technology within Organisations", *International Journal of Information Management* 16(1): pp. 65–74.

Dickinson, B. (1997) "Knowing that the Project Clothes have no Emperor", *Knowledge and Process Management* 4(4): pp. 261–267.

Doty, P. and Erdelez, S, (2002) "Information Micro-practices in Texas Rural Courts: Methods and Issues for E-government", *Government Information Quarterly*, 19: pp. 369–387.

Drejer, A. (1996) "Frameworks for the Management of Technology: Towards a Contingent Approach", *Journal of Technology Analysis Management* 8(1): pp. 9–20.

Drury, D. (1983) "An Empirical Assessment of the Stages of OP Growth", *MIS Quarterly* 7(2): pp. 59–70.

Earl, M. (1989) "Approaches to Strategic Information Systems Planning Experience in Twenty-One United Kingdom Companies", in *Proceedings Tenth Annual International Conference on Information Systems*, pp. 271–277.

*Economist* (1999) "A Survey of the Business and the Internet: The net imperative", *The Economist*, June, UK.

Egan, J. (1998) *Rethinking Construction: The Report of the Construction Task Force*, London: Department of the Environment, Transport and Regions.

El Emam, K., Drouin, J. and Melo, W. (1998) "SPICE: The Theory and Practice of Software Process Improvement and Capability Determination", *IEEE Computer Society*.

Farbey, B., Land, F. and Targett, D. (1992) "Appraising Investments in IT", *Journal of Information Technology* 7: pp. 109–122.

Farbey, B., Targett, D. and Land, F. (1994) "The Great Benefit Hunt", *European Management Journal* 12(3): pp. 270–279.

Feeny, D. and Willcocks, L. (1998) "Core IS Capabilities for Exploiting Information Technology,", *Sloan Management Review* (Spring): pp. 9–21.

Feeny, D.F., Edwards, B.R. and Simpson, K.M (1992) "Understanding the CIO/CEO Relationship", *MIS Quarterly* 16(4): pp. 435–448.

Fountain, J. (2001) *Building the Virtual State: Information Technology and Institutional Change*, Washington DC: Brookings Institution.

Galliers, R. and Sutherland, A. (1991) "Information Systems Management and Strategy Formulation: The Stages of Growth Model Revisited", *Journal of Information Systems* 1(2): pp. 89–114.

Garrity, E. and Sanders, G. (1998a) *Information Systems Success Measurement*, Hershy, PA: Idea Group.

Garrity, E. and Sanders, G. (1998b) "Dimensions of Information Systems Success", in E. Garrity and G. Sanders (eds) *Information Systems Success Measurement*, Hershy, PA: Idea Group.

Gasser, L. (1986) "The Integration of Computing and Routine Work", *ACM Transactions on Office Information Systems* 4(3): pp. 205–225.

Gatian, A. (1994) "Is User Satisfaction A Valid Measure of System Effectiveness?" *Information and Management* 26(3): pp. 119–131.

Geber, B. (1994) "A Clean Break for Education at IBM", *Journal of Training* 31(2): pp. 33–36.

Geletkanycz, M. (1997) "The Salience of Culture's Consequences: the Effects of Cultural Values on Top Executive Commitment to the Status Quo", *Strategic Management Journal* 18(8): pp. 615–634.

General Accounting Office (2001) "Electronic Government: Challenges Must be Addressed with Effective Leadership and Management", GAO-01-959T, 11 July, pp. 1–2.

Giaglis, G.M., Paul, R.J. and O'Keefe, R.M. (1999) "Research Note: Integrating Business and Network Simulation Models for IT Investment Evaluation", *Logistics Information Management* 12(1/2): pp. 108–117.

Gibbs, W. (1994) "Software's Chronic Crisis", *Scientific American* 271(3): pp. 86–95.

Ginzberg, M. (1981) "Early Diagnosis of MIS Implementation Failure: Promising Results and Unanswered Questions", *Management Science* 27(4): pp. 459–478.

Gottschalk, P. and Khandelwal, V. (2002) "Global Comparison of Stages of Growth Based on Critical Success Factors", *Journal of Global Information Management* 10(2): pp. 40–49.

Goulding, J.S. (2000) "GAPP-T: A Generic IT Training Model for Construction", PhD dissertation, School of Construction and Property Management, University of Salford, Greater Manchester.

Goulding, J.S. and Alshawi, M. (1997) "Construction Business Strategies: A Synergetic Alliance of Corporate Vision, Information Technology, and Training Strategies", in *Proceedings The First International Conference on Construction Industry Development*, Singapore.

Goulding, J.S. and Alshawi, M. (1999) "Generic IT Training: A Process Protocol Model", in *Proceedings The Durability of Building Materials and Components Conference 8*, Vancouver, Canada.

Goulding, J.S. and Alshawi, M. (2002) "Generic IT Training: A Process Protocol Model for Construction", *Journal of Construction Management and Economics* 20(6): pp. 493–505.

Griggs, F.E.J. (1997) "To Be or Not To Be – Ethical That IS", *Journal of Professional Issues in Engineering Education and Practice* 123(2): pp. 82–89.

Grindley, K. (1992) "Information Systems Issues Facing Senior Executives: The Culture Gap", *Journal of Strategic Information Systems* 1(2): pp. 57–62.

Grover, V., Jeong, S. and Segars, A. (1996) "Information Systems Effectiveness: the Construct Space and Patterns of Application", *Information and Management* 31: pp. 177–191.

Grover, V., Teng, J. and Fiedler, K. (1993) "Information Technology Enabled Business Process Redesign: An Integrated Planning Framework", *Omega: The International Journal of Management Science* 21(4): pp. 433–447.

Gunning, J.G. (1996) "Practical Change Management in Construction", in *Proceedings CIB W89 Beijing International Conference*, Beijing, China.

Hall, M.I. (1997) "Motivation: Dispelling Some Management Myths – An Analytical Critique", *Journal of Construction Education* 1(3): pp. 99–108.

Hall, T.J. (1997) "The Quality Systems Manual: The Definitive Guide to the ISO 9000 Family and TickIT", *Journal of Operational Research Society*, 48(1): pp. 105.

Hamel, G. (1991) "Competition for Competence and Inter-Partner Learning within International Strategic Alliances", *Strategic Management Journal* 12: pp. 83–103.

Hamilton, S. and Chervany, N. (1981) "Evaluating Information Systems Effectiveness – Part I: Comparing Evaluation Approaches", *MIS Quarterly* 5(3): pp. 55–69.

Hammer, M. (1990) "Reengineering Work: Don't Automate, Oliberate", *Harvard Business Review* (July/August): pp. 104–112.

Hammer, M. and Champy, J. (1993) *Reengineering the Corporation: A Manifesto for Business Revolution*, New York: Harper Business.

Hammer, M. and Stanton, S. (1995) *The Re-engineering Revolution*, New York: Harper Collins.

Harmon, P. (2004) "Evaluating an Organisation's Business Process Maturity", *Business Process Trends* 2(3): pp. 1–11.

Harrison, E.F. and Pelletier, M.A. (2000) "Levels of Strategic Decision Success", *Management Decision Journal* 38(2): pp. 107–117.

Harvey, C. and Denton, J. (1999) "To Come of Age: The Antecedents of Organizational Learning", *Journal of Management Studies* 36(5): pp. 897–918.

Heng, L. (1996) "The Role of IT Manager in Construction Process Reengineering", *Building Research and Information* 24(2): pp. 124–130.

Hibbard, J. and Carillo, K.M. (1998) "Knowledge Revolution – Getting Employees to Share what they Know is no longer a Technology Challenge – it's a Corporate Culture Challenge", *Information Week*, available online at http://www.informationweek.com/663/63iuknw.htm (accessed April 2007).

Hiller, J. and Belanger, F. (2001) *Privacy Strategies for Electronic Government*, Arlington, VA: PricewaterHouseCooper Endowment for Business of the Government.

Hinterhuber, H.H. (1995) "Business Process Management: The European Approach", *Business Change & Re-engineering* 2(4): pp. 63–73.

Hirschheim, R., Earl, M., Feeny, D. and Lockett, M. (1988) "An Exploration into the Management of Information Systems Function: Key Issues and an Evolutionary Model", in *Proceedings Information Technology Management for Productivity and Strategic Advantage, IFIP TC-8 Open Conference*, Singapore.

Hitt, L. and Brynjolfsson, E. (1996) "Productivity and Consumer Welfare: Three Different Measures of Information Technology's Value", *MIS Quarterly* 20(2): pp. 121–143.

Hoos, I. (1960) "When Computers Take Over the Office", *Harvard Business Review* 38(4): pp. 102–112.

Huber, G.P. (1991) "Organisational Learning: The Contributing Processes and the Literatures", *Journal of Organisation Science* 2(1): pp. 88–115.

Humphrey, W. (1995) *A Discipline for Software Engineering*. (SEI Series in Software Engineering) New York: Addison-Wesley.

Humphrey, W. (1998a) "Three Dimensions of Process Improvement – Part I: Process Maturity. CROSSTALK", *The Journal of Defence Software Engineering* (February): pp. 14–17.

Humphrey, W. (1998b) "Three Dimensions of Process Improvement – Part II: The Personal Process. CROSSTALK", *The Journal of Defence Software Egineering* (March): pp. 13–15.

Humphrey, W. (1998c) "Three Dimensions of Process Improvement – Part III: The Team Process. CROSSTALK", *The Journal of Defence Software Engineering* (April): pp. 14–17.

Humphrey, W., Lovelace, M. and Hoppes, R. (1999) *Introduction to the Team Software Process*. (SEI Series in Software Engineering), New York: Addison-Wesley.

IDPM (2002) "Definition of IT", available online at http://www.man.ac.uk/idpm (accessed 5 April 2002).

Ingram, P. and Baum, J.A.C. (1997) "Opportunity and Constraint: Organizations' Learning from the Operating and Competitive Experience of Industries", *Strategic Management Journal* 18: pp. 75–98.

ISO/IEC (1995a) *ISO/IEC 9126–1: Information Technology – Software Quality Characterisitc and Metrics. Version 3.2.*

Ives, B., Olson, M.H. and Baroudi, J.J. (1983) "The Measurement of User Information Satisfaction", *Communications of the ACM* 26(10): pp. 785–793.

Jaeger, P.T. and Thompson, K.M. (2003) "e-Government around the World: Lessons, Challenges, and Future Directions", *Government Information Quarterly* 20: pp. 389–394.

Johansson, H., McHugh, P., Pendlebury, J. and Wheeler, W. (1993) *Business Process Re-engineering: Breaking Point Strategies for Market Dominance*, Chichester: John Wiley & Sons.

Johnson, G. (1992) "Managing Strategic Change – Strategy, Culture and Action", *Journal of Long Range Planning* 25(1): pp. 28–36.

Joia, L.A. (2000) "Measuring Intangible Corporate Assets – Linking Business Strategy with Intellectual Capital", *Journal of Intellectual Capital* 1(1): pp. 68–84.

Kagioglou, M., Cooper, R., Aouad, G., Hinks, J., Sexton, M. and Sheath, D. (1998) *Final Report: Process Protocol*, University of Salford, Manchester.

Kangas, K. (1999) "Competency and Capabilities Based Competition and the Role of Information Technology: The Case of Trading by a Finland-based Firm to Russia", *Journal of Information Technology Cases and Applications* 1(2): pp. 4–22.

Kaplan, R.S. and Norton, D.P. (1992) "The Balanced Scorecard – Measures That Drive Performance", *Harvard Business Review* (Jan/Feb): pp. 71–79.

Kaplan, R.S. and Norton, D.P. (1996) "Using the Balanced Scorecard as a Strategic Management System", *Harvard Business Review* 74(1): pp. 75–85.

Karim, K., Marosszeky, M., Devalence, G. and Miller, R.M. (1997) "Practice and Priorities for Benchmarking in Australian Construction Industry", in *Proceedings 1st International Conference on Construction Industry Development: Building the Future Together*, University of Singapore, Singapore, pp. 332–339.

Kaylor, C., Deshazo, R., Van Eck, D. (2001) "Gauging e-Government: A Report on Implementing Services Among American Cities", *Government Information Quarterly* 18: pp. 293–307.

Kennedy, C. (1994) "Re-engineering: The Human Costs and Benefits", *Long Range Planning* 27(5): pp. 64–72.

Kessels, J. and Harrison, R. (1998) "External Consistency: The Key to Success in Management Development Programmes", *Journal of Management Learning* 29(1): pp. 39–68.

King, J. and Schrems, E. (1978) "Cost Benefit Analysis in IS Development and Operation", *Computing Surveys* 7(2): pp. 19–34.

King, J.L. and Kraemer, K.L. (1984) "Evolution and Organisational Information Systems: An Assessment of Nolan's Stage Model", *Communications of the ACM* 27(5): pp. 466–485.

Klein, J., Gee, D. and Jones, H. (1998) "Analysing Clusters of Skills in R&D – Core Competencies, Metaphors, Visualisation, and the Role of IT", *R&D Management Journal* 28(1): pp. 37–42.

Klenke, K. (1994) "Information Technologies as Drivers of Emergent Organizational Forms: A Leadership Perspective", in R. Baskerville, S. Smithson, O. Ngwenyama and J. DeGross (eds) *Transforming Organisations with Information Technology*, Amsterdam: Elsevier Science, pp. 323–341.

Klimko, G. (2001) "Knowledge Management and Maturity Models: Building Common Understanding", *Proceedings of the 2nd European Conference on Knowledge Management*, Bled School of Management, Bled, Slovenia, 8–9 November, pp 269–278.

KPMG (1990) *Runaway Computer Systems*, London: KMPG Peat Marwick McLintock.

KPMG (2000) *Knowledge Management Survey Report*, London: KMPG Consulting Publications.

Koch, C. (2002) "The Emergence of Second Generation Knowledge Management in Engineering Consulting", International Council of Research and Innovation in Building and Construction, CIB W78.

Koch, C. (2003) "Knowledge Management in Consulting Engineering: Joining IT and Human Resources to Support the Production of Knowledge", *Engineering, Construction and Architecture Management* 10(6): pp. 391–401.

Koch, C., Slater, D. and Baatz, E. (1999) *The ABC's of ERP*, London: CIO.

Kriebel, C. and Raviv, A. (1980) "An Economics Approach to Modelling the Productivity of Computer Systems", *Management Science* 26(3): pp. 297–311.

Krogt, F. and Warmerdam, J. (1997) "Training in Different Types of Organisations: Differences and Dynamics in the Organisation of Learning at Work", *International Journal of Human Resource Management* 8(1): pp. 87–105.

Kumaraswamy, M. (1997) "Improving Industry Performance Through Integrated Training Programs", *Journal of Professional Issues in Engineering Education and Practice* 123(3): pp. 93–97.

Kuvaja, P., Simila, J., Krzanik, L., Bicego, A., Kocj, G. and Saukkonen, S. (1994) *Software Process Assessment and Improvement: The BOOTSTRAP Approach*. London: Blackwell.

Lawton, G. (2001) "Knowledge Management: Ready for Prime Time", *Computer* 34(2): pp. 12–14.

Layne, K. and Lee, J. (2001) "Developing Fully Functional E-government: A Four Stage Model", *Government Information Quarterly* 18: pp. 122–136.

Leat, M.J. and Lovell, M.J. (1997) "Training Needs Analysis: Weaknesses in the Conventional Approach", *Journal of European Industrial Training* 24(4): pp. 143–153.

Leavitt, H.J. and Whistler, T.L (1958) "Management in the 1980s", *Harvard Business Review* (November–December): pp. 41–48.

Lee, S.F. and Sai On Ko, A. (2000) "Building Balanced Scorecard with SWOT Analysis, and Implementing Sun Tzu's 'The Art of Business Management Strategies' on QFD Methodology", *Managerial Auditing Journal* 15(1/2): pp. 68–76.

Legge, K., Clegg, C. and Kemp, N. (1991) *Case Studies in Information Technology, People and Organisations*, Oxford: NCC Blackwell.

Leonard-Barton, D. (1992) "Core Capabilities and Core Rigidities: A Paradox in Managing New Product Development", *Strategic Management Journal* 13: pp. 111–125.

Leonard-Barton (1995) *Wellsprings of knowledge*. Harvard Business School Press. Boston, MA.

Lientz, B. and L. Larson (2004) *Manage IT as a Business: How to Achieve Alignment and Add Value to the Company*, Burlington, MA: Elsevier.

Ly, E. (1997) *Distributed Java Applets for Project Management over the Web*, Industry Report. IEEE.

Lynch, T. and Gregor, S. (2001) "User Involvement in DSS Development: Patterns of

Influence and System Impact", in *Proceedings 6th International Society for Decision Support Systems Conference* (ISDSS2001) Brunel University, West London, pp. 207–217.

McCuen, R. (1998) "Balancing Corporate and Personal Values", *Journal of Management in Engineering* 14(2): pp. 40–44.

McKinsey (2005) *Does IT Improve Performance?* Chart Focus Newsletter, Member Edn, *The McKinsey Quarterly*.

Maier, R. and Remus, U. (2003) "Implementing Process-oriented Knowledge Management Strategies", *Journal of Knowledge Management* 7(4): pp. 62–74.

Majchrzak, A. (1988) *The Human Side of Factory Automation*, San Francisco: Jossey-Bass.

Maloney, W.F. (1997) "Strategic Planning for Human Resource Management in Construction", *Journal of Management in Engineering* 13(3): pp. 49–56.

Marsh, L. and Flanagan, R. (2000) "Measuring the Costs and Benefits of Information Technology in Construction", *Engineering and Architectural Management* 7(4): pp. 423–435.

Maslow, A.H. (1970) *Motivation and Personality*, 2nd edn, New York: Harper and Row.

Mata, F.J., Fuerst, W.L and Barney, J.B. (1995) "Information Technology and Sustained Competitive Advantage: A Resource-Based Analysis", *MIS Quarterly* 8(19): pp. 487–505.

May, M. (1999) "Developing Management Competencies for Fast-Changing Organisations", *Career Development International Journal* 4(6): pp. 336–339.

Mayo, A. (2000) "The Role of Employee Development in the Growth of Intellectual Capital", *Journal of Personnel Review* 29(4): pp. 521–533.

Meso, P. and Smith, R. (2000) "A Resource-based View of Organizational Knowledge Management Systems", *Journal of Knowledge Management* 4(3): pp. 224–234.

Meso, P., Troutt, M.D. and Rudnicka, J. (2002) "A Review of Naturalistic Decision Making Research with Some Implications for Knowledge Management", *Journal of Knowledge Management* 6(1): pp. 63–73.

Meyer, A.D. (1982) "How Ideas Supplant Formal Structures and Shape Responses to Environments", *Journal of Management* 19(1): pp. 45–61.

Meyer, M.H. and Utterback, J.M. (1993) "The Product Family and the Dynamics of Core Capability", *Sloan Management Review* (Spring): pp. 29–47.

Michalski, G.V. and Cousins, J.B. (2000) "Differences in Stakeholder Perceptions about Training Evaluation: A Concept Mapping/ Pattern Matching Investigation", *Journal of Evaluation and Program Planning* 23(2): pp. 211–230.

Miles, D.W.J. and Neale, R.H. (1997) "Patterns in Diversity: An International Training Systems Typology", in *Proceedings 1st International Conference on Construction Industry Development: Building the Future Together*, University of Singapore, Singapore, pp. 142–152.

Mintzberg, H. and Quinn, J.B. (1991) *The Strategy Process: Concepts, Contexts, Cases*, Englewood Cliffs, NJ: Prentice Hall.

Mockler, R. and Dologite, D. (1995) "Easing Information Technology Across Cultural Boundaries – A Contingency Perspective", *International Journal of Computer Applications in Technology* 8(3–4): pp. 145–162.

Mohamed, S. and Tilley, P.A. (1997) "Benchmarking for Best Practice in Construction" in *Proceedings 1st International Conference on Construction Industry Development: Building the Future Together*, University of Singapore, Singapore, pp. 420–427.

Moingeon, B., Ramanantsoa, B., Me'tais, E. and Orton, J.D. (1998) "Another Look at Strategy–Structure Relationships: The Resource-based View", *European Management Journal* 16(3): pp. 298–304.

Moon, J. (2002) "The Evolution of E-Government among Municipalities: Rhetoric or Reality?" *Public Administration Review*, July/August, 62(4): pp 424–433.

Mulcahy, J.F. (1990) "Management of the Building Firm", in *Proceedings CIB 90, Joint Symposium on Building Economics and Construction Management*, Sydney, pp. 11–21.

Mumford, E. (1995) "Creative Chaos or Constructive Change: Business Process Re-engineering versus Socio-technical Design", in G. Burke and J. Peppard (eds) *Examining Business Process Re-engineering: Current Perspectives and Research Directions*, New York: Kogan Page, pp. 192–216.

Munshi, J. (1996) "Framework for MIS Effectiveness", Proceedings of Academy of Business Administration, International Conference, Athens, Greece, pp. 26–32.

Musso, J., Weare, C. and Hale, M. (2000) "Designing Web Technologies for Local Governance Reform: Good Management or Good Democracy", *Political Communication* 17(1): pp 1–19.

Myers, B., Kappelman, L. and Prybutok, V. (1998) "A Comprehensive Model for Assessing the Quality and Productivity of the Information Systems Function: Towards a Theory for Information Systems Assessment", in E. Garrity and G. Sanders (eds) *Information Systems Success Measurement*, Hershy, PA: Idea Group.

Naoum, S. and Hackman, S. (1996) "Do Site Managers and Head Office Perceive Productivity Factors Differently?" *Engineering Construction and Architecture Management* 3(1/2): pp. 147–160.

Neary, S.J. and Yeomans, D. (1996) "Education and Practice – A One Way Street", in *Proceedings CIB W89 Beijing International Conference*, Beijing, China.

Nolan, R. (1979) "Managing the Crisis in Data Processing", *Harvard Business Review* (March/April): pp. 115–126.

Nonaka, I. (1991) "The Knowledge Creating Company", *Harvard Business Review* 6(8): pp. 96–104.

Nonaka, I. and Takeuchi, H. (1995) *The Knowledge Creating Company – How Japanese Companies Create the Dynamics of Innovations*, Oxford : Oxford University Press.

Obaid, A. (2004) "Model for a Successful Implementation of Knowledge Management in Engineering Organisations", PhD Thesis, Research Institute of the Built Environment, University of Salford, UK.

Office of Technology Assessment (1984) *Computerized Manufacturing Automation*, Library of Congress no. 84–601053, Government Print Office, Washington DC.

Olomolaiye, P.O., Jayawardane, A.K.W. and Harris, F.C. (1998) *Construction Productivity Management*, London: Addison Wesley Longman.

Ovenden, T. (1994) "Business Process Re-engineering: Definitely Worth Considering", *The TQM Magazine* 6(3): pp. 56–61.

Page, D., Williams, P. and Boyd, D. (1993) *Report of the Enquiry into the London Ambulance Service*, London: South West Thames Regional Health Authority.

Pascale, A.T. and Athos, A.G. (1981) *The Art of the Japanese Management*, New York: Warner.

Paulk, M. (1995) "How ISO 9001 Compare to CMM", *IEEE Software*, 12(1): pp. 74–83.

Paulk, M., Curtis, B., Chrissis, M. and Weber, C. (1993) *Capability Maturity Model for*

*Software, Version 1.1*. Technical Report CMU/ SEI-93-TR-024. Software Engineering Institute/Carnegie Mellon University.

Paulson, T. (1995) *Paulson on Change*, (Griffin's Distilled Wisdom Series) Griffin Publishing Inc.

Pemberton, J.D. and Stonehouse, G.H. (2000) "Organizational Learning and Knowledge Assets – An Essential Partnership", *The Learning Organization* 7(4): pp. 184–193.

Penrose, E.T. (1959) *The Theory of the Growth of the Firm*, New York :John Wiley & Sons.

Peppard, J. (1995) "Management Challenges in Information Systems", *Journal of Information Technology* 10: pp. 127–130.

Peppard, J. and Ward, J. (2004) "Beyond Strategic Information Systems: Towards an IS Capability", *The Journal of Strategic Information Systems* 13(2): pp. 167–194.

Perkowski, J.C. (1988) "Technical Trends in the E&C Business: The Next 10 Years", *Journal of Construction Management and Engineering* 114(4): pp. 565–576.

Philip, G., Gopalakrishnan, M. and Mawalkar, S.R. (1995) "Technology Management and Information Technology Strategy – Preliminary Results of an Empirical Study of Canadian Organisations", *International Journal of Information Management* 15(4): pp. 303–315.

Pickrell, S. and Garnett, N. (1996) "Generic Benchmarking in Construction", in *Proceedings CIB W89 Beijing International Conference*, Beijing, China.

Porter, M.E. (1985) *Competitive Advantage*, New York: Free Press.

Porter, M.E. and Millar, V.E (1985) "How Information Gives You Competitive Advantage", *Harvard Business Review* 63 (July/ August): pp. 149–160.

Powell, T. and Dent-Micallef, A. (1997) "Information Technology as Competitive Advantage: The Role of Human, Business and Technology Resources", *Strategic Management Journal* 18(5): pp. 375–405.

Prahalad, C.K. and Hamel, G. (1990) "The Core Competence of the Corporation", *Harvard Business Review* 68 (May–June): pp. 79–91.

Price, R.K., Solomatine, D.P. and Velicov, S. (2000) "Internet-based Computing and Knowledge Management for Engineering Services", in *Proceedings 4th Conference on Hydroinformatics*, Iowa City.

Quinn, B., Anderson, P. and Finkelstein, S. (1996) "Managing Professional Intellect, Making the Most of the Best", *Harvard Business Review* (March–April): pp. 71–80.

Radcliffe, R. (1982) *Investment: Concepts, Analysis, Strategy*, Glenview, III: Scott Foreman.

Raghuram, S. (1994) "Linking Staffing and Training Practices with Business strategy: A Theoretical Perspective", *Human Resource Development Quarterly* 5(3): pp. 237–251.

Ramsay, W. (1989) "Business Objectives and Strategy", in P.M. Hillebrandt and Cannon, J. (eds) *The Management of Construction Firms: Aspects of Theory*, London: Macmillan.

Reiblein, S. and Symons, A. (1997) "SPI: I Can't Get No Satisfaction – Directing Process Improvements to Meet Business Needs", *Software Quality Journal* 6(2): pp. 89–98.

Roach, S. (1987) *America's Technology Dilemma: A Profile of Information Economy*, Morgan Stanely Special Economic Study.

Roach, S. (1991) "Service under Siege: The Restructing Imperative", *Harvard Business Review* 39(2): pp. 82.

Robson, W. (1997) *Strategic Management and Information Systems*, London: Pitman Publishing.

Rockart, J.F., Earl, M.J. and Ross, J.W. (1996) "Eight Imperatives for the New IT Organization", *Sloan Management Review* 38: pp. 43(13)

Rohm, C. (1992/1993) "The Principal Insures a Better Future by Re-engineering its Individual Insurance Department", *National Productivity Review* 12(1): pp. 55–64.

Ross, J. (1998) "IT infrastructure management", in *Proceedings 98 Annual Conference of Information Systems Management*, Centre for Research in Information Management, London Business School, London.

Ross, J.W., Beath, C.M. and Goodhue, D.L. (1996) "Develop Long-Term Competitiveness Through IT Assets", *Sloan Management Review* (Fall): pp. 31–42.

Rus, I. and Lindvall, S. (2002) "Knowledge Management in Software Engineering", *IEEE Software* (May/ June): pp. 26–38.

Sainter, P., Oldham, K., Larkin, A., Murton, A. and Brimble, R. (2000) "Product Knowledge Management within Knowledge-based Engineering Systems", in *Proceedings ASME 2000 Design Engineering Technical Conference*, Baltimore, MD.

Salah, Y. (2003) "IS/IT Success and Evaluation: A General Practitioner Model", PhD Thesis, Research Institute for the Built Environment, University of Salford, UK.

Sarshar, M., Haigh R., Finnemore M., Aouad G., Barrett P., Baldry D., Sexton M. (2000) "SPICE: A Business Process Diagnostics Tool for Construction Projects", *Engineering Construction & Architectural Management* 7(3): pp. 241–266.

Scanlin, J. (1998) "The Internet as an Enabler of the Bell Atlantic Project Office", *Project Management Journal* (June): pp. 6–7.

Scarborough, J. (1997) "Making the Matrix Matter: Strategic Information Systems in Financial Services", *Journal of Management Studies* 34(2): pp. 171–190.

Scarborough, H., Swan, J. and Preston, J. (1999) *Knowledge Management: A Review of the Literature*, London: Institute of Personal Development.

Schnitt, D. (1993) "Re-engineering the Organisation using Information Technology", *Journal of Systems Management*: pp. 14–20.

Schott, H., Rose, T. and Schlick, C. (2000) "Process Knowledge Management in Concurrent Engineering", in *Proceedings 2nd European Systems Engineering Conference (EuSEC)*, Munich.

Seddon, P. (1997) "A Respecification and Extension of the DeLone and McLean Model of IS Success", *Information Systems Research* 8(3): pp. 240.

Segers, A. and Grover, V. (1998) "Strategic Information Systems Planning Success: An Investigation of the Construct and Its Measurement", *MIS Quarterly* 22(2): pp. 139–163.

SEI (1999) *Capability Maturity Model Integration (CMMI) Project*, Software Engineering Institute, Carnegie Mellon University.

SEI (2001) *People Capability Maturity Model (Version 2.0)*, Software Engineering Institute, Carnegie Mellon University.

Senge, P.M. (1990) *The Fifth Discipline. The Art and practice of the Learning Organisation*, London: Century Business.

Shankar, R., Singh, M.D., Gupta, A. and Narain, R. (2003) "Strategic Planning for Knowledge Management Implementation in Engineering Firms", *Work Study* 52(4): pp. 190–200.

Shirazi, B., Langford, D.A. and Rowlinson, S.M. (1996) "Organizational Structures in the Construction Industry", *Construction Management and Economics* 14(3): pp. 199–212.

Simpson, W. (1987) *New Techniques in Software Project Management*, New York: John Wiley & Sons.

Skyrme, D. (1997) *Creating the Knowledge Base Business*, London: Business Intelligence Ltd.

Sleezer, C. (1993) "Training Needs Assessment at Work: A Dynamic Process", *Human Resource Development Quarterly* 4(3): pp. 247–264.

Soibelman, L. and Caldas, C. (2000) "Information Logistics for Construction Design Team Collaboration", in *Proceedings 8th International Conference on Computing in Civil and Building Engineering* (viii–ICCCBE) Stanford, CA.

Solingen, R. and Berghout, E. (1999) *The Goal/Question/Metric Method: A Practical Guide for Quality Improvement of Software Development*, New York: McGraw-Hill.

Somogyi, E.K. and Galliers, R.D. (1987) "Applied Information Technology: from Data Processing to Strategic Information Systems", *Journal of Information Technology* 2(1): pp. 30–41.

Spender, J.C. (1996) "Making Knowledge the Basis of a Dynamic Theory of the Firm", *Strategic Management Journal* 17: pp. 45–62.

Spendolini, M. (1992) "The Benchmarking Process", *Journal of Compensation and Benefits Review* 24(5): pp. 21–29.

Standish Group (1995) *The Chaos Report*, West Yarmouth, MA: Standish Group.

Standish Group (2001) *The Chaos Report*, West Yarmouth, MA: Standish Group.

Stockman, S. and Norris, M. (1991) "Engineering Approaches to Software Development in the 90s, in N. Fenton and B. Littlewoods (eds) *Software Reliability and Metrics*, Amsterdam: Elsevier Applied Science.

Strassman, P. (1997) "Will Big Spending on Computers Guarantee Profitability?" *Datamation*, available online at http://www.strassmann.com (accessed 4 April 2002).

Subramanian, A. and Lacity, M.C. (1997) "Managing Client/ Server Implementations: Today's Technology, Yesterday's Lessons", *Journal of Information Technology* 12(3): pp. 169–186.

Swanson, E. (1974) "Management Information Systems: Appreciation and Involvement", *Management Science*, 21(2): pp 178–188.

Tan, R.R. (1996) "Information Technology and Perceived Competitive Advantage: An Empirical Consulting Firms in Taiwan", *Construction Management and Economics* 14(3): pp. 227–240.

Tapscott, D. and Caston, A. (1993) *Paradigm Shift: The New Promise of Information Technology*, New York: McGraw-Hill.

Taylor, M. (2003) *E-BC Strategic Plan, Performance Measures*, Version 3, State of British Columbia.

Taylor-Cummings, A. (1998) "Bridging the User-IS Gap: A Study of Major Information Systems Projects", *Journal of Information Technology* 13(1): pp. 29–54.

TechTarget (2002) "Definition of IT", available online at http://search390.techtarget-.com/sDefinition/0,,sid10_gci214023,00.html (accessed5 April 2002).

Teece, D.J. (2000) "Stragies for Managing Knowledge Assests: The Role of Firm Structure and Industrial Context, *Long Range Planning* 33: pp 35–54.

Towers, S. (1996) "Re-engineering: Middle Managers are the Key Asset", *Management Services* 40(12): pp. 17–18.

Trillium (1996) *Trillium Model: For Telecom Product Development and Support Process Capability*, Model Issue 3.2. Bell Canada.

Van Daal, B., De Haas, M. and Weggeman, M. (1998) "The Knowledge Matrix: A

Participatory Method for Individual Knowledge Gap Determination", *Knowledge and Process Management* 5(4): pp. 255–263.

Venegas, P. and Alarcón, L. (1997) "Selecting Long-Term Strategies for Construction Firms", *Journal of Construction Engineering and Management* 123(4): pp. 388–398.

Venkatraman, N. (1991) "IT-induced Business Re-configuration", in M. Scott-Morton (ed.) *The Corporation of the 1990s*, Oxford: Oxford University Press, pp. 121–158.

Verona, G. (1999) "A Resource-based View of Product Development", *Academy of Management Review* 24(1): pp. 132–141.

Walsh, J.P. and Ungson, G.R. (1991) "Organisational Memory", *Academy of Management Review* 16(1): pp. 57–91.

Walshman, G. (1993) "Reading the Organisation: Metaphors and Information Management", *Journal of Information Systems* 3: pp. 33–46.

Ward, J. and Griffiths, P. (1997) *Strategic Planning for Information Systems*, Chichester: John Wiley & Sons.

Warszawski, A. (1996) "Strategic Planning in Construction Companies", *Journal of Construction Engineering and Management* 122(2): pp. 133–140.

Watson, R.T. (1990) "Influences on the IS Manager's Perceptions of Key Issues: Information Scanning and the Relationship with the CEO", *MIS Quarterly* (14): pp. 217–231.

Weill, P. and Broadbent, M. (1999) "Four Views of IT Infra-structure", in L. Willcocks and S. Lester (eds) *Beyond the IT Productivity Paradox*, Chichester: John Wiley & Sons.

Weill, P., Broadbent, M. and St. Clair, D. (1994) "IT Value and the Role of IT Infrastructure Investments", CISR WP; no 268. Working paper, Center for Information Systems Research, Sloan School of Management, Massachusetts Institute of Technology.

Wijeyesekera, D. (1997) "Training and Education for Construction Industry Development", in *Proceedings The 1st International Conference on Construction Industry Development: Building the Future Together*, University of Singapore, Singapore, pp. 69–76.

Willcocks, L. and Lester, S. (1996) "Beyond the IT Productivity Paradox", *European Management Journal* 14(3): pp. 279–290.

Willcocks, L. and Smith, G. (1994) *IT-enabled Business Process Re-engineering: From Theory to Practice*, Oxford: Oxford University Working Papers.

Wimmer, M.A. (2002) "A European Perspective Towards Online One-stop Government: the Egov Project", *Electronic Commerce Research and Applications* 1: pp. 92–103.

Woodroof, J. and Kasper, G. (1998) "A Conceptual Development of Process and Outcome User Satisfaction", in E. Garrity and G. Sanders (eds) *Information Systems Success Measurement*, Hershy, PA: Idea Group.

Xia, W. and G. Lee (2005) "Complexity of Information Systems Development Projects: Conceptualization and Measurement Development", *Journal of Management Information Systems* 22(1): pp. 45–83.

Zahran, S. (1997) *Software Process Improvement, Practical Guidelines for Business Success*, New York: Addison-Wesley.

# Index

Note: *italic* page numbers denote references to figures/tables

Milton Keynes UK
Ingram Content Group UK Ltd.
UKHW040445071024
449327UK00020B/997